Math Momentum in Science Centers

Math Momentum
in Science Centers

Edited by Jan Mokros

Authors:
    Jamie Bell
    Jan Mokros
    Ricardo Nemirovsky
    Andee Rubin
    Tracey Wright

 TERC

National
Science
Foundation

TERC

Association of
Science-
Technology
Centers

This book is based upon work of the "Math Momentum in Science Centers" project, funded by the National Science Foundation's Informal Science Education Program under Grant Number 0229782. Any opinions, findings, and conclusions or recommendations expressed in this book are those of the authors and do not necessarily reflect the views of the National Science Foundation (NSF).

The project was led by TERC, in conjunction with the Association of Science-Technology Centers (ASTC) and thirteen science centers and aquariums nationwide, including Buffalo Museum of Science; Children's Museum of Houston; Fort Worth Museum of Science and History; Lawrence Hall of Science; Miami Museum of Science; Museum of Science, Boston; New England Aquarium; New Jersey Academy for Aquatic Sciences; Museum of Life and Science, Durham, North Carolina; Oregon Museum of Science and Industry; Science Museum of Minnesota; Sciencenter, Ithaca, New York; and Saint Louis Sciencenter.

TERC
2067 Massachusetts Avenue
Cambridge, MA 02140
www.terc.edu

ISBN-10: 1-929877-10-2
ISBN-13: 978-1-929877-10-2

Design and Production by Jim Mafchir
Editing by Cinny Green and Jill Pellarin
Drawings unless otherwise noted by Maureen Burdock
Printed in Canada

Contents

Foreword

George Hein, Professor Emeritus, Lesley University

Science centers are a growth industry. The American Association of Science-Technology Centers (ASTC), founded in 1973, currently has 540 member institutions in 40 countries. A parallel European-based organization, the European Collaborative for Science, Industry and Technology Exhibitions (ECSITE), founded about 15 years later, already has 385 members in 30 different European countries. Even more recently, the Asia Pacific Network of Science and Technology Centers (APSTC), less than 10 years old, claims 20 members in the region and a few in North America and Europe; the older Indian governmental National Council of Science Museums (NCSM), founded in 1978, operates 27 science centers under a single administrative structure.

Science centers, as distinguished from more traditional science museums that often emphasized objects and the presentation of science subject matter in an orderly manner, perceive themselves primarily as places where visitors can learn through experience and take delight in the activities of science, as well as gather information about science. Not infrequently, these centers combine traditional natural science with various applications—technology, science policy, and industrial or historic uses of science—or incorporate other topics in their exhibitions and programs, such as art, culture, or social science.

Perhaps surprisingly, mathematics as a specific focus or even as a supplement to the content of science centers has been only marginally explored, despite the essential relationship between quantification and scientific investigation. The impressive exhibition Mathematica: The World of Numbers and Beyond, created by the Eames workshop in 1961, still tours; and one of several early versions is still installed at the Museum of Science, Boston. Various exhibitions about mathematics also appear occasionally. For example, Mirrors of the Mind, a Swiss exhibition that exploits geometry was at Heureka in Finland in 2005–2006. And Handling Calculus, a novel effort to translate mathematical functions into three-dimensional interactive exhibitions, based on research by Ricardo Nemirovsky at TERC (described in this volume), was recently at the Science Museum of Minnesota. Certainly there are others elsewhere, some included in this volume. But considering the thousands of ingenious exhibitions developed since the first science centers opened in 1969, those focused on mathematics, or even those that incorporate means for visitors to use mathematics, are shockingly limited.

Much of the power of scientific explanation, especially in the physical sciences that provide the majority of exhibitions in science centers, is derived from the ability to derive quantitative correlations, rather than only qualitative ones. Newton titled his famous treatise *Principia Mathematica Philosophiae Naturalis* (*The Mathematical Principles of Natural Philosophy*), emphasizing the quantitative basis for his insights. Yet science centers have opted primarily to illustrate principles and provide experiences that do not require careful measurement, repeated trials, or making quantitative comparisons. The authors of this volume point out that "the three mathematical areas where science centers have the greatest potential to incorporate math are data, measurement, and the mathematics of change (patterns, algebra, and calculus)." All are core mathematical topics.

The reasons for limiting mathematics in science centers are not hard to find. On the whole, visitors do not have the time or patience, not to mention the learned skills, needed to do or appreciate the nature of the role of mathematics in science. Even adults do not commonly understand, for example, the need for repeated measurements (and how to arrive at a reasonable average from repeated measures) or how to assess experimental error; nor do they commonly have strategies to recognize and then minimize systematic errors. Children are often eager to experiment and willing to repeat "neat" phenomena, but they seldom record results, even in those rare instances when the museum provides them the means to do so (Gottfried, 1981). Designing interactive exhibitions that allow fail-proof visitor activity—and are robust enough to withstand constant use—often requires restricting variables so severely that meaningful inquiry is limited to fairly obvious results that can be appreciated qualitatively.

But other factors have led to a reexamination of the role of mathematics in science centers in recent years. Foremost has been the increased national concern in the past two decades with mathematical competence among students and the entire population, as well as the increased realization of the important role that informal educational settings (museums, after-school, day care, etc.) can play in providing essential experiences for acquiring knowledge. Simultaneously, science centers have been forging stronger links with schools, and educators within these institutions are becoming increasingly involved in exhibition development, and launching more comprehensive educational programs utilizing their resources. In 1998, YouthALIVE!, a powerful ASTC program that brought thousands of secondary-school students from disadvantaged backgrounds into science centers, sponsored a conference for science-center professionals on the topic of mathematics in science centers. An NSF-funded report commissioned by ASTC (Anderson, 2001), based on case studies of five science centers, illustrated the potential for mathematics-focused exhibits and programs. This was followed by the Math Momentum in Science Centers project, again funded by NSF, and carried out jointly by TERC and ASTC, resulting in this volume, among other products.

The Math Momentum project was, on the one hand, a fairly straightforward professional development program for science-center professionals. ASTC has sponsored other such programs; for example, one on accessible practices for a three-year period, 2000–2002, involved six lead institutions that hosted professional development workshops for other museums. However, the Math Momentum project also incorporated some unique features that are evident in this publication.

First, Math Momentum had a dual mandate to promote mathematics education through science-center programs and exhibits, and to address issues of equity, another long-term goal of both ASTC and TERC. The book devotes a whole chapter to the important "quest for mathematical equity." The authors assert, "Instead of thinking about math as a filter, it could be thought about as a 'pipeline' that could bring as many people as possible into as many areas of math as possible." A belief in the educability of all learners is a necessary, even if by itself insufficient, step towards achieving equity in any educational activity.

The authors also emphasize a point that is important for all educators to accept: Science-center staff can learn much about the abilities and skills of their visitors by listening and by observing what the visitors bring to a mathematical situation, before attempting to instruct.

Second, the activities described and promoted have been extended beyond serving individual visitors to science centers. The book also includes a discussion of the educational roles available to parents and how mathematics can be incorporated into outreach activities and issues of social justice. In short, the manuscript can serve as a guide to science-center professionals for expanding their use of mathematics into any and all aspects of their professional work.

Jan Mokros and her coauthors have also woven into the narrative many specific examples of how the science centers that participated in this project incorporated mathematics into a wide range of programs and exhibits. These examples, for the most part, don't involve elaborate exhibitions and complex technology, but commendably, use many relatively simple activities and materials that still engage the participant in sophisticated mathematical issues. Wonderful activities of this sort are available today, based on curriculum materials developed by many groups (including TERC's own *Investigations*). Their mathematical richness and mental sophistication have been tested in schools and documented by a host of publications that describe how children and adults can grapple with these examples, utilizing various means to engage with the particular inquiries. Mathematics can be—as a number of recent NSF-funded projects have emphasized—a field of inquiry offering an opportunity for experimentation and individual initiative, not simply a subject that requires applying memorized algorithms.

Finally, the project and this book emphasize the need for science-center staff not only to develop math-related activities for their visitors, but also to engage in math activities themselves. "The most important ingredient in getting started with math is the willingness to look more closely at the math around you, and to examine how people (including yourself) engage in this math." We all benefit if we experience what we propose to teach.

We can only hope that the science centers of the future will be *mathematical* science centers that entice visitors to engage in math activities with the same energy and enthusiasm that many now demonstrate when they explore science exhibitions and programs.

References

Anderson, A. 2001. *Mathematics in Science Centers*. Washington, DC: American Association of Science-Technology Centers.

Gottfried, J. 1981. Do children learn on field trips? *Curator: The Museum Journal* 23(3):165–74.

Acknowledgments

The Math Momentum in Science Centers project was a deeply collaborative effort involving TERC, the Association of Science-Technology Centers (ASTC), and thirteen science centers and aquariums nationwide, including Buffalo Museum of Science; Children's Museum of Houston; Fort Worth Museum of Science and History; Lawrence Hall of Science; Miami Museum of Science; Museum of Science, Boston; New England Aquarium; New Jersey Academy for Aquatic Sciences; Museum of Life and Science, Durham, NC; Oregon Museum of Science and Industry; Science Museum of Minnesota; Sciencenter, Ithaca, NY; and Saint Louis Sciencenter. We heartily thank all of these centers for their commitment to "mathematizing" their work and their creativity and productivity in making math part of what they do on a daily basis. A sampling of the projects that these science centers are engaged in is sprinkled throughout the book; a compilation of a representative project from each center is in the appendix.

The National Science Foundation (NSF) funded the project, and we are especially grateful to Informal Science Education program officers Julie Johnson and Sylvia James for their advice and commitment to the role of math in science centers.

TERC, which served as the prime contractor for the Math Momentum in Science Centers project, is a nonprofit organization devoted to math and science education. TERC's work includes research, curriculum design, professional development, and the development of programs for informal learning environments. We are grateful for the organizational support TERC provided to the project throughout its lifespan.

The Association of Science-Technology Centers served as a

major subcontractor to the Math Momentum Project. At ASTC, Bonnie Van Dorn, DeAnna Banks Beane, and Jacquelyn Lowery built upon their pioneering work in bringing math and science centers together. Throughout the project, they ensured that science centers throughout the country were paying attention to math and to the needs of a broad range of science-center audiences. They did this through math sessions and CEO breakfasts at ASTC conferences, a lively math Special Interest Group, an issue of ASTC *Dimensions*, and tireless networking. They played critical roles in the project's institutes and workshops, as well as in its outreach efforts. In large part due to Jackie's and DeAnna's efforts, over 350 science-center staff attended Math Momentum workshops.

The project's advisors contributed in substantive ways to our institutes and helped the project specify the many important meanings of "mathematizing." Advisors included math educators, family math education specialists, and researchers and practitioners in informal math education. Thanks to advisors Andrea Anderson, Ana Becerra, Marta Civil, Francena Cummings, José Franco, George Hein, Cliff Konold, Macenje Mazoka, Judit Moschkovich, J. Newlin, William Tate, Virginia Thompson, and Carol Valenta.

Our evaluators, consultants, and colleagues to whom we owe a special note of thanks include Dennis Bartels (science-center wisdom), Christina Cupples (administrative support), Cecilia Garibay (evaluation), Lori Lambertson (program and activity development at the Exploratorium), Jim Mafchir of Western Edge Publications, Valerie Martin (graphics and permissions), Jill Pellarin (editor), David Smith (videography), and Samuel Taylor (formative evaluation).

Most of the TERC staff who served on the project are also chapter authors and are listed below. But the TERC staff person to whom we owe the biggest debt of gratitude is Bronwyn Low. Bronwyn was the conductor of Math Momentum from start to finish and worked wonders with our large, talented, and sometimes disorganized orchestra. It was Bronwyn who planned in large part the institutes, workshops, and meetings; created the project's Web site; organized the financial and contractual work; kept us on track; and created the outline for this book. As a student of folklore and anthropology, Bronwyn brought perspectives that deeply enriched the math education work of this project.

The Development of the Math Momentum Book

The chapters are authored by people on the TERC team who worked on the project over the course of three years. They are as follows:

- Jamie Bell, Director, Centre of Learning, Petrosains: The Discovery Centre, Kuala Lumpur, Malaysia

- Jan Mokros, TERC Principal Scientist and Principal Investigator of Math Momentum in Science Centers

- Ricardo Nemirovsky, Professor, Department of Mathematics and Statistics, San Diego State University

- Andee Rubin, TERC Principal Scientist

- Tracey Wright, TERC Senior Research Associate

Thanks to the reviewers of chapters, all of whom were involved in the Math Momentum project, including DeAnna Beane, Jayme Cellitioci, Marilyn Johnson, Cheryl Juarez, Kathy Krafft, Robert "Chip" Lindsay, Diane Miller, Keith Ostfeld, Maija Sedzielarz, Harriette Stevens, and Loren Stolow.

In addition to these authors and reviewers, several of our science-center partners "put meat on the bones" of the chapters by writing sections for "Voice from the Field" to illustrate the very real challenges and opportunities that arise when math becomes a focus.

CHAPTER 1

Why Math in Science Centers?

Jan Mokros

For many people, math is like kitchen clean-up: a necessary, rote, and occasionally satisfying experience, but one that is rarely relished. In school, math is interesting to younger students, but only up to a point. Elementary-school students typically enjoy math, but interest fades during middle school and wanes even more as students progress through the gatekeeper courses of Algebra I and II. Though some people continue to love math, one of the leading reasons students give for dropping out of high school is the difficulty or boredom they experienced in math (Viadero, 2005).

Unfortunately, those students who lose interest in math usually stop taking math courses, and as a result they miss out on many career opportunities, particularly in science and social science. In work, in the community, and at home, people who cannot do math are seriously restricted. They are not prepared to contribute fully as citizens because they cannot make good sense of data on issues ranging from local recycling to drilling for oil in the Arctic. Their personal lives may be impoverished because they don't have the skills to make optimal decisions about finances, health care insurance, and retirement. For example, people who are uncomfortable with math are more risk averse, less apt to make good investment decisions, and less financially prepared for retirement (Burns, 1998, p.140). They often are forced to pay others to do tasks that should be well within their reach, such as financial planning, tax prep, or even cooking the right amount for a party.

Although most people recognize its importance in their lives, math is rarely thought about as an important and interesting pursuit in and of itself. In fact, the idea of math for fun seems ludicrous to most of us. What would it look like? Marilyn Burns, one of a rare breed of educators who finds "everyday" math fascinating, issues a call for action: "Let's get the cereal makers involved and put on the backs of cereal boxes engaging ways to think about math so families can start their mornings with breakfast and math. Let's get all the late-night TV show hosts to include at least one good math problem in their opening monologues... Let's get movie theaters to devote a bit of preshow screen time to some good math. Maybe we need a Broadway show: *Math—The Musical...* Let's get the ball rolling" (Burns, 1998, p. 143).

And who better than science centers to get the ball rolling? After all, they provide plenty of activities involving balls, ramps— and potentially, math! If centers thought about themselves as clubhouses for doing math, what might this scenario look like? How would science centers be different from the way they are now? On a journey to an imaginary science center that has many playing fields for doing math, here are some activities that might be going on.

First stop: Step up to the reaction-time exhibit to see how fast you can press a button when you see a light or hear a sound. Each time you try, you see your time on a screen and get a small printout of it, along with a few words congratulating you on getting faster. Your pals get similar feedback, color-coded for each person. When you've all tried your luck, press a button to see graphs showing how your reaction time, as well as your friends', changed over a few trials. Who was the fastest? Who showed the most improvement? How did your reaction time compare with those of people older and younger than you? Look at the "Record Reaction Times" chart next to the exhibit (culled from the experiences of previous visitors) to figure out where you stand.

Second stop: Kids—and their families—love rolling balls down ramps or, more aptly, *racing* balls down ramps to see whose gets to the end first. But how do you measure which ramp is fastest? With a few simple measurement tools, including a tape measure to determine the lengths of different ramps and large stopwatches to record times accurately, visitors can make meaningful mathematical comparisons. Your ball may have reached the end first, but if it had less distance to travel, did you really win? How can you tell for sure?

Third stop: Don't miss the calculus exhibit. *Calculus?* So many people are in line that you'll have to wait your turn to move the life-size kinetic snowboarder through a series of moguls and turns, while figuring out where she gains and loses momentum and where her speed sends her rocketing along the course or careening out of control. To be successful, you'll need to think a lot about speed and acceleration. It will help to talk through your strategy with other visitors.

Next on the agenda is the live animal show featuring the resident sharks. As usual, visitors have a few misconceptions about sharks, one of which relates to their voracious appetites. During the show, the explainer asks, "How much do you think you'd eat if you were this 200-pound shark?" She lays out many quarter-pound rubber hamburgers on the table, and visitors make their estimates. You find out, by weighing and reasoning, that sharks have small appetites; a small child eats the same amount daily as a 200-pound shark.

On your way to lunch, you are captivated by a real-life fantasy figure blowing bubbles. She is catching the imprint of the bubbles on paper and has asked a group of kids to figure out who can blow the biggest bubble. Given the ephemeral quality of bubbles, you can't judge by just looking at the bubbles themselves. Some bubble-print circles are clearly bigger than others, but what does that tell us about the size of the bubbles? How can we measure? Captain Subtracta is equipped with an array of tools, including string, and she skillfully encourages children to suggest viable ways of using these tools. You watch as a child measures around the circumference of two circles with two different pieces of string, clips a paper clip onto the end points of the two strings, compares them, and finally, with pride, announces the bubble winner. Other children, impressed with this method, shout, "I want to try!" and line up for a chance to measure bubbles.

This imaginary journey is not fictional; all of these experiences are part of exhibits at real science centers, although the journey is a composite of scenes from a number of different centers.[1] These centers are part of an NSF-funded project, Math Momentum in Science Centers, which is building momentum for math by linking math with science, broadening visitors' notions about math, and making math accessible to all kinds of visitors, including those who they think are not interested in math. The thirteen centers[2] are making math a priority because they want to engage their visitors in important mathematics that supports science, learning, and their institutional missions. They have found that exciting math is a way to accomplish serious educational and institutional goals. Collectively, centers participating in the

[1] The examples are culled from Children's Museum of Houston; Museum of Science, Boston; New Jersey Academy for Aquatic Sciences; Science Museum of Minnesota; and Sciencenter, Ithaca, NY.
[2] Partners include Buffalo Museum of Science; Children's Museum of Houston; Fort Worth Museum of Science and History; Lawrence Hall of Science at University of California, Berkeley; Miami Museum of Science, Inc.; Museum of Science, Boston; New England Aquarium; New Jersey Academy for Aquatic Sciences; Museum of Life and Science, Durham, NC; Oregon Museum of Science and Industry; Science Museum of Minnesota; Sciencenter, Ithaca, NY; and Saint Louis Sciencenter.

project have made the following realizations:

✓ Math can help people see what real scientists do as they collect and analyze data, and thereby can help people understand the interplay of science and math. Establishing the relationship between math and science is part of the mission of many science centers.

✓ Science centers broaden people's perspectives about the nature and content of science—and can do the same for math. Math is much more than arithmetic and more than a subject to be studied in school.

✓ Math comes alive in science centers through hands-on inquiry and can increase visitors' excitement about learning. Exciting hands-on learning is a goal of virtually all science centers.

✓ Math is a critical tool for achieving equity. The gap in math test scores between whites and people of color is substantial, and science centers serve a vital role in closing this gap by bringing everyone to the mathematical playing field.

✓ Exhibits that incorporate mathematical challenges often serve as a powerful means of engaging visitors and keeping them engaged.

✓ Science centers take pride in bridging school learning with real-world learning and can show the myriad uses of math in the world (thus addressing the question "When are we ever going to have to use this?").

✓ Providing math experiences for families gives parents tools they need and value to help their children learn, while fostering quality family time.

✓ Providing math experiences for outreach programs gives community-based organizations tools to help their youth learn in an engaging way.

Where's the Math?

We have visited science centers around the country, working with staff to find the math hidden beneath the surface of activities, exhibits, and programs. We've looked for opportunities to easily incorporate math into existing programs, as well as programs that are being designed. And we've found that there are certain kinds of mathematical opportunities that are pervasive in science centers.

Finding this math means stepping back for a moment and reconsidering what math is. The National Council of Teachers of Mathematics (NCTM), through its curriculum *Principles and Standards* (NCTM, 2000), reminds us that math is far more than number operations. NCTM advocates for learning the kind of math that schools often neglect, including the mathematics of data, geometry, measurement, and pattern-finding. Interestingly, these undertaught areas of math are just the places where science centers offer a multitude of opportunities. The three mathematical areas where science centers have the greatest potential to incorporate math are data, measurement, and the mathematics of change (patterns, algebra, and calculus).

Data

All of science involves data collection, analysis, and interpretation, but these critical mathematical processes are not yet very visible in science centers. This is probably due to the popular perception that data is of little interest to visitors. Working with data does not, at first glance, seem to be the kind of hands-on experience that visitors enjoy. Visitors certainly enjoy exploring the phenomena, but do they really want to deal with the mathematical "drudgery" involved in collecting and working with data? Interestingly, scientists themselves often love doing the mathematical work, as it is central to answering their questions and solving the mysteries they are untangling. Why couldn't this be the case for visitors as well?

For scientists, doing an experiment without collecting and savoring data is like eating a meal without a main course. Data are the meat of any investigation, and science centers owe it to visitors to provide them with a main course. Otherwise, it may seem as if science is all activity, with no analysis and no unraveling of a mystery.

There is another more practical reason for "doing data" in science centers: Visitors love to find out about themselves, compare themselves with others, and answer their own questions about a phenomenon. Working with data empowers them to address these questions and to do so in a customized way. When working with data, visitors can address questions that are meaningful to them.

For example, think again about the reaction-time exhibit, one that is ubiquitous in science centers. There are many scientific questions involved in studying reaction time, including questions that visitors themselves often ask, such as these:

✓ Is reaction time faster in response to a sound or to a visual cue like a light?
✓ Are reaction times of older people slower? Is this an area

where children are faster than their parents?

✓ Does reaction time improve with experience? How much experience does it take?

To answer any of these questions, one must collect and examine data. Relying on memory ("Who was faster—Aunt Juana, brother, or me?") is no way to do science. To help visitors address scientific questions, feedback must be provided in the form of immediate, easy-to-use data. If you have a data trail and can see that Aunt Juana's time was 0.21 seconds, brother's was 0.34 seconds, and yours was 0.29 seconds, you will be able to answer the question "Who in our group was fastest?" Of course, you will need to know something about what these decimal numbers mean, and to understand that 0.34 is not the fastest reaction time, just because it is the largest number. This is where well-designed support materials come in. We will see more about how this works in chapter 4.

Data at school and in science centers: Because data are the essence of all scientific and social science experiments, educators are being encouraged to put more emphasis on the study of data in math classes throughout elementary and high school. Educators are increasingly recognizing that learning to tell the story of a set of data is a critical mathematical skill. Telling these stories accurately involves, at a minimum, seeing how actual numbers in a data set are clustered or spread out; examining the middle, average, or typical values in a data set; determining what the unusual or outlying pieces of data mean; and comparing two or more sets of data to see how they differ. All of these skills involve mathematical reasoning.

Ideally, building this kind of reasoning starts at a young age by motivating children to collect and use data to answer questions that have some importance to them. In math class, elementary teachers are encouraged to involve children in collecting real data (not limited to surveys of favorite ice cream flavors!) and analyzing the results to make decisions, such as which books the library might purchase, which activities will be set up during recess, or how tall the chairs should be in the lunchroom, given the heights of students in the school.

Teachers see the importance of working with data in math class, but often do not have enough class time to do justice to this approach. They would like students to examine data sets, and they wish that it could happen more often. This is where science centers enter the picture. They can provide dynamic interactive data sets that allow visitors to contribute their own data on a phenomenon like reaction time, see how their data fit into the bigger picture of data from many other individuals, and address questions about who is fastest, what the record time is, and whether practice makes a difference.

Measurement

As is the case with data, most work in science involves measurement. Results of almost any experiment need to be measured, and the range of measures is astounding. Scientists measure linear distances (from nanometers to astronomical units), area, volume, capacity, mass, and time. They construct and use derived measures, such as perceived exertion on a treadmill test. Scientists need to choose appropriate measures, determine which measuring tools are right for the job, figure out which attributes need to be measured (no easy task, as it turns out), and measure the phenomenon accurately. They also need to figure out why measurements vary, and how much variability is acceptable and how much is just background noise.

Measurement work is more difficult than it appears to be at first glance, particularly when the attributes being measured are too small or large to grapple with in a concrete way. The emerging work of science centers on nanotechnology is addressing complex issues of how visitors, especially younger ones, make sense of measurement in worlds that are too small to be seen. Certainly, understanding small worlds will be made easier by having a grounding in the processes of measurement and the ways in which measurement tools are used.

In everyday life, as in scientific endeavors, measurement skills are critical. We cannot function without being able to figure out how long, how far, and how much. Scheduling our days depends on understanding the measurement of time; making grocery purchases necessitates an understanding of how much family members eat; and getting around depends on an understanding of measurements of distance and space.

Given the fact that measurement is a foundational skill for science and for real life, it is disappointing that U.S. children perform so poorly on formal assessments of measurement. In international tests, U.S. students rank near the bottom in terms of their measurement skills and understanding (Programme for International Student Assessment, 2000). Moreover, when all of the areas of math achievement are examined for U.S. students,

measurement is an area where there is consistently poor performance. (Clements and Bright, 2003).

What do students find difficult?

In fourth grade, less than 25% of students can correctly
identify the length of the line segment in a task like the one shown
below (Lindquist and Kouba, 1989)

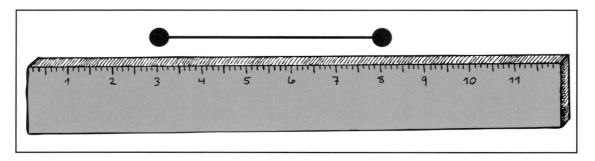

How long is this line segment?

One additional cause for concern is that the biggest gaps in
math performance between children of color and Caucasian children are in the area of measurement (Lubienski, 2003). There is
little agreement about why this is the case; however, educators
agree that more must be done to narrow this gap. The simple, critical act of giving children more opportunities to measure is an
excellent starting place to help level the playing field.

Measurement at school and in science centers: According to the
National Council of Teachers of Mathematics, "It is unlikely that
children can gain a deep understanding of measurement without
handling materials, making comparisons physically, and measuring with tools" (NCTM, 2000, p. 44). In other words, in order to
learn to measure, children must practice measuring on a regular basis.

In working with teachers on measurement activities, we have
found that they face several challenges. First, there isn't enough
class time to give students enough real experience in measuring.
Because measurement skills, unlike other areas of math, can be
taught and practiced at home, teachers may pare down ambitious
mathematics agendas by cutting out skills that can be learned elsewhere; measurement is one of these skills. Second, learning to
measure involves tools, noise, and movement. Tools such as scales,
stopwatches, and tape measures are expensive for schools, and
there is often not enough space for a classroom full of students to

practice measuring. Additionally, because measuring often involves a high amount of noise, it may be frowned upon in some school settings.

When it comes to practicing measuring, as well as measuring for a real purpose, science centers offer unparalleled potential. They are busy, active places where visitors are encouraged to use a variety of tools. There are many hands-on educational programs that could easily involve purposeful measurement and make use of measuring tools. And noise is not an issue!

Yet how often do centers involve visitors in measuring? How often do they strategically place measuring tapes near exhibits where distance matters or put timers near those where speed is of the essence? Science centers could provide more opportunities for practicing measurement. A variety of ways of making headway with measurement are described in chapters 4, 7, and 8.

Patterns, algebra, and calculus

Math is used by scientists of all kinds to describe patterns, movement, growth, and changes over time, as well as to make clear representations of how these changes occur. Often scientists do this by using graphs and tables and, increasingly, by using spatial representations such as maps that show several variables at once. Scientists use principles of algebra and calculus as they study everything from the spread of disease to the paleontology of the Grand Canyon to charting the course of hurricanes. If science centers want to show visitors how these kinds of science are done, they must help visitors see the underlying mathematics.

When people think about algebra and calculus, many react with boredom, disengagement, and even fear. But these reactions are often based on how the topics have been taught in disembodied ways, not on an understanding of the wide range of fascinating applications of these ideas. Most of us think of algebra and calculus as more relevant for an older audience. Often we do not think about how younger children make sense of these ideas. Yet educational researchers are finding that even very young children make use of the mathematics underlying algebra and calculus. For example, researchers at Tufts University are working with eight- to ten-year-olds to study how they make comparisons, reason about general relationships (rather than specific numbers), and make use of tables to search for patterns (Schliemann, Carraher, and Brizuela, 2003). Examining numerical relationships is at the core of the work, and it is important for children to begin this work early.

Where are algebra and calculus found in science centers? Just about everywhere. For example, almost every science center has activities or exhibits showing pendulums, the motion of planets, plant growth, and how sound waves travel and change. Many exhibits and hands-on activities in physical science involve motion. Consider the ubiquitous car-and-ramp activities in science centers. Visitors are often asked to determine which ramp will be fastest, one that is a straight downhill course or one that has a series of bumps. They are encouraged to race two cars or balls down ramps to see which one is fastest. And there are a lot of questions along the course of these routes: When do the cars speed up or slow down? Does the car on the straight downhill course gain speed as it moves along? What about the car on the bumpy course—where is its speed the fastest? Do you lose more speed going uphill than you gain going downhill? Tools exist for slowing down, representing, and examining motion, and these tools can help visitors explore the mathematics of motion. In chapter 5, "Exhibits," we will see how eagerly visitors move within the physics and mathematics of motion as they address compelling questions.

Patterns, algebra, and calculus at school and in science centers:
In school, the need to get students involved with the mathematics of change has never been more apparent. Both algebra and calculus are known as gatekeeper subjects. If you do well in these areas, you pass Go, get into college, and often have your choice of high-paying jobs and careers. If you bypass these subjects or do poorly in them, your choices are more limited. The question of who takes algebra and calculus courses is a major equity issue. Historically in the United States, Caucasian and Asian American children—especially boys—have been far more likely to complete high-school algebra and college calculus. Welcome changes are occurring (see chapter 3), and "algebra for all" is a goal that is being realized in communities across the country (Moses and Cobb, 2001). Girls are now achieving on par with boys in high-school math classes, though not as much progress has been made in achieving racial equality in access to mathematics.

Unfortunately, schools and teachers encounter an uphill battle as they try to encourage students to enroll in algebra, precalculus, and calculus courses. The courses often have bad reputations among students and are thought of as distasteful medicine that must be taken. Students have precious few opportunities to experience the relevance, beauty, and power of these disciplines, nor do they often see how math helps us make sense of the world.

One could argue that helping people make sense of the physical world is one of the major goals of science centers. If this is true, then algebra and calculus are the most potent tools that centers have as they engage visitors in understanding the world. In science centers, there are no artificial boundaries between disciplines that are kept separate in schools. Algebra and calculus allow you to examine how plants grow, how musical patterns are shaped, how pole-vaulters take off, and the rate at which bird flu could spread.

Inquiry in Science—and in Math

Most readers are quite familiar with the approach that engages visitors in doing science, rather than just reading about it or watching others do it. Hands-on, or inquiry-based, science has become the buzzword in the mission statements of many science centers. It is a given for many, if not most exhibits. But what about an inquiry approach to math? Throughout this book, we argue that inquiry in math is in many ways parallel to inquiry in science and demands the active hands-on engagement of visitors. In fact, our assumption is that you can't learn math unless you roll up your sleeves and actively do it. This principle applies not only to visitors, but also to education staff, exhibit developers, youth programs staff, volunteers, and administrators. Math inquiry works in the same way that exhibits themselves are constructed—with a lot of group work, some great beginning questions, multiple approaches, and a lot of starts and stops along the way.

Inquiry in math is in some ways different from inquiry in science, and the differences will be touched upon throughout this book. One of the differences, in many cases, is that less "stuff" is involved in math. Also, the starting point isn't as often an awesome physical phenomenon. It may seem like a lot of math is cerebral, rather than hands-on. However, if looked at in a fresh way, math involves using all of one's senses: Visual beauty can emerge from exploring fractals; the sounds and patterns of music can be heard in new ways; and the motion of downhill skiing can be captured using a combination of graphs and narrative.

Math is alive, useful, delightful, and practical. It is a tool for understanding science, a means for negotiating daily life, and a way of thinking that is beautiful in and of itself. It is time for science centers to convey these new messages about math!

References

Burns, M. 1998. *Math: Facing an American Phobia*. Sausalito, CA: Math Solutions Publications.

Clements, D. and G. Bright. 2003. Eds. *Learning and Teaching Measurement: 2003 Yearbook*. Reston, VA: National Council of Teachers of Mathematics, Inc.

Lindquist, M. and V. Kouba. 1989. Geometry. In *Results from the Fourth Mathematics Asssessment of the National Assessment of Educational Progress*. M. Lindquist. Ed. Reston, VA: National Council of Teachers of Mathematics, Inc.:44-54.

Lubienski, S. 2003. Is our teaching measuring up? Race-, SES-, and gender-related gaps in measurement achievement. In Clements and Bright. Eds. *Learning and Teaching Measurement: 2003 Yearbook*. Reston, VA: National Council of Teachers of Mathematics, Inc.:282–292.

Moses, R. and C. Cobb. 2001. Radical Equations: *Civil Rights from Mississippi to the Algebra Project*. Boston, MA: Beacon Press.

National Council of Teachers of Mathematics, Inc. 2000. *Principles and Standards for School Mathematics*. Reston, VA: NCTM.

Programme for International Student Assessment. 2000. *Measuring Student Knowledge and Skills: The PISA 2000 Assessment of Reading, Mathematical, and Scientific Literacy*. Organization for Economic Co-operation and Development Publishing. Downloaded from www.PISA.oecd.org.

Schliemann, A. et al. 2006. *Bringing Out the Algebraic Character of Arithmetic: From Children's Ideas to Classroom Practice*. Hillsdale, NJ: Lawrence Erlbaum Associates.

Viadero, D. 2005. Math emerges as big hurdle for teenagers. *Education Week* 24(28):1-16.

CHAPTER 2

Getting Started:
Math Momentum Begins with You!

Jan Mokros

Science-center staff wear a variety of hats but have few opportunities to wear the hat of learner. The available professional development is often brief, scattered, or packed into conferences where there are many competing sessions. In this chapter, we argue that the starting place for "mathematizing" programs or exhibits is immersing *yourself* in at least four types of learning: learning math, doing math together with your colleagues, examining how visitors learn math, and finding out how scientists and other content experts in your center actually use math in their work. There is an even more important goal that encompasses all four of these types of learning—the goal of seeing math differently and seeing it in myriad ways throughout one's institution and throughout the course of everyday life.

Why is your own learning the essential foundation for incorporating math into exhibits and programs? In math and science education in schools, there are many indications that learning won't take hold deeply unless there is active inquiry. In science centers, "active prolonged engagement" (Humphrey and Gutwill, 2005) is becoming a key criterion of designing exhibits and programs. Mathematical inquiry and prolonged engagement are as essential to math as they are to science, and as essential for staff as they are for visitors.

Inquiry and prolonged engagement: essential to math and to science: Science centers are known for their skill in promoting active, hands-on inquiry in science. For example, a good exhibit developer or program specialist can take a principle of physics and, with levers, pulleys, or balls and ramps, translate it into an experience that gives visitors an immediate, multisensory experience of the physics principle. But this type of work doesn't often happen when the topic is mathematics. Given most of the experiences people had as students, they often find it difficult to see that mathematics involves active questioning, constructing, manipulation of objects, and experimenting, or that it can involve significant hands-on work and prolonged engagement. It is also hard to imagine that mathematics is more than analytical problem-solving or that it involves many of the multiple intelligences, including spatial, logical, naturalistic, and kinesthetic intelligences, that are part of Gardner's evolving theory on that subject (Gardner, 1999). Yet it is precisely this multisensory, active math that is best suited to science centers.

Inquiry and prolonged engagement: equally important for staff and for visitors: Busy professionals are often tempted to bypass the experience of doing the science—or the math—that underpins the exhibits or programs they are developing. But common sense dictates that if the activity isn't interesting or meaningful to its designer, it's unlikely to engage the visitor. Similarly, if professionals learn little science or math from the experiences they are creating—or if they can't make sense of the underlying concepts—it is unlikely that visitors will have powerful learning experiences. For these reasons, it's important for staff to take time on a regular basis to experience mathematics directly, do mathematics in conjunction with colleagues, observe the math that visitors engage in, and have mathematical conversations with scientists and other experts at their centers. This is a challenging process, and a necessary one.

Because it is difficult to envision what this mathematics might look like, we begin by inviting you to engage in a math problem that shows something of the nature of inquiry-based, hands-on math. As you engage with the problem, think about how the nature of inquiry is similar to (or different from) that involved in science challenges you've encountered.

Make copies of each of the strange cakes below. Then divide each of them into two equal halves. The two halves need not have the same shape. Use whatever tools you like, and feel free to cut, fold, color in, etc. Note that there is not one right answer. Several different solutions work with each of the figures.

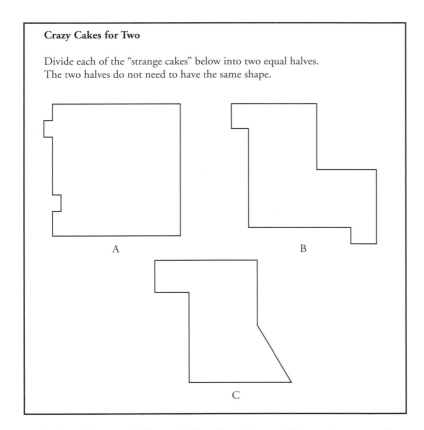

Crazy Cakes for Two

Divide each of the "strange cakes" below into two equal halves.
The two halves do not need to have the same shape.

A

B

C

Tierney, C., M. Ogonowski, A. Rubin, and S. J. Russell. 1998. Different shapes, equal pieces. Investigations in Number, Data, and Space. Menlo Park, CA: Dale Seymour Publications.

Principles of Good Professional Development

Principle 1: First, do math yourself

As you've seen from doing Crazy Cakes, math doesn't have to mean doing sets of problems from textbooks. It can and should mean grappling with problems that have multiple paths to solution, where there may be more than one solution, and where you don't need to recall a particular formula or bit of arcane mathematical knowledge to solve a problem. For example, in dividing the oddly shaped cakes into equal halves, you may have begun by thinking about what "half" means in the context of these odd shapes. You may have thought about how each cake could be divided in half fairly. Trying to create equal-size pieces is something we've all experienced—and probably had numerous arguments over! Of course, there's a twist to this problem: You were given nonsymmetrical pieces to prompt you to think a little harder about the question "How can one make equal-size pieces when the pieces aren't *shaped* equally?" The important mathematical idea underlying this question is *equivalence*; two pieces can cover the same area, but be very different in shape.

What strategies did you use to divide the shapes into halves? Many people think about the cakes visually and simply "see" the tradeoffs that can be made to make the two halves equivalent in area. Other people do a conscious tit-for-tat strategy and adjust smaller pieces within the shape to make both halves equivalent. Still others make use of the relationships between triangles and squares, because squares can easily be divided into triangles.

We've provided some of the typical solutions others have come up with for these problems at the end of the chapter to give you a sense of the multiple ways they can be addressed. Your solutions may be different, but what is important is that you be able to demonstrate or prove that they work. One way to do this is to copy the shapes on a piece of paper, then cut and paste to see whether the areas of the two halves are truly the same. Doing this kind of informal proof is critical, as it involves making sense of the mathematics. If you can't convince yourself or someone else that your solution works, an important part of the mathematics is missing.

The Crazy Cakes problem was intentionally constructed to promote deep thinking, and though it was designed for a math curriculum, it's the type of problem that could be part of a science center program. It is meant to pull you into math, whatever your background and attitude about the subject, and to make you wonder what half means. You may want to reflect for a moment on whether and how the experience worked for you.

✓ Did you try different approaches or strategies to divide each shape, or did you give up?
✓ Did you have a couple of ways to try the problem, or did you feel it was hopeless?
✓ Did solving the problem make you feel that you were learning or constructing something for yourself, or did you feel you needed an external authority (in this case, the page of answers) to validate your work?

In the process of doing math, you will learn about the characteristics of math problems that promote learning and engagement. In fact, we'd argue that the less you enjoy doing math yourself, the more accurate your barometer will be for discerning whether the math involved in a specific activity will be engaging to visitors.

Voice from the Field[1]

Integrating Math into the Culture of Children's Museums

Keith Ostfeld, Director of Visitor Experience, Children's Museum of Houston

This interview was conducted in July 2005 by Jan Mokros.
Quotations are not exact.

Why did you decide to involve staff in professional development that centers on math?
Keith: A lot of people don't realize that in children's museums, a substantial amount of math happens. Sometimes, staff are surprised by this, and they themselves may be phobic about math. Their last math experience in school may have been with algebra, so math often doesn't make sense to them. We felt as an institution that it was important for us to be committed to doing math with visitors. That meant we had to involve staff in doing math in a significant way.
We wanted to get math fully integrated into our culture, and the way to do this is to bring staff in and show them that math is not what they think it is—that it's fun and that they can do it.

What kinds of experiences did you make available?
Keith: We have a variety of experiences, some on-site and others that are workshops people can take

off-site. We have brought in math educators to do workshops that would give staff a sense of how different math is from what they thought—that it's not just about doing worksheets. Regularly, we do workshops that center on the math kits we've developed. Floor staff and educators find these workshops really useful because they center on a particular kit that these people will be using with visitors. We have also sent people to Math Solutions workshops to Developing Mathematical Ideas workshops, which involve at least a week's worth of time. Here, people can delve into the math more deeply and learn a great deal. We were lucky to have grant money to let us do some of this work.

Who has been involved in these experiences and how did you convince them to go?
Keith: Our director is really committed to math, and she made it clear that this was going to be a priority. To some extent it was required, but

[1] In this section—and others throughout the book—entitled "Voice from the Field," a Math Momentum partner describes direct experience with integrating math into a science-center program.

people were paid for their time and they usually jumped at the opportunity. To do your work well, you need to understand more about math, so that was a big motivator. For our floor staff, we made a lot of opportunities available and also offered food. Almost everyone, except a few administrators, has been involved in some form of professional development in math. It's definitely expected of staff.

What kind of impact did you notice once a critical mass of staff had been involved in these experiences?

Keith: After one or two trainings, you'd go up to the Math Cart (which provides mediated math activities for visitors) and see that staff are much more enthusiastic about a particular kit, that they have more ideas about how to work with visitors around it, and that they are taking it to the next level. I also see staff talking to each other about math in the staff room. They are recognizing where the math is. They even talk about how they approach the kits differently with different people. That's really exciting! Some of our educators are tweaking the kits as a result of the professional development they've done. Once, a floor staff member came up to me after a workshop and said she had been thinking about ways of integrating math into a different activity. That was really satisfying, to find people thinking about generalizing the math to other contexts. We find, more generally, that staff understand math is not what they thought it was, that it is much more active and based on solving different kinds of problems.

What advice do you have for others about doing professional development in math?

Keith: First, don't rush. It takes time for things to change. You have to have a long-term commitment to it. Second, don't do it all yourself. You may want to do some training yourself, particularly around your own programs or kits, but staff also need to get more opportunities to learn about math from a broader perspective. Bring in a math educator to work with as a consultant—and involve the whole institution in a workshop that she or he does. Third, make math a major part of your culture. This involves a process of enculturation where everyone buys into it. Staff development can't be a one-shot thing. Offer opportunities on a regular basis so that new people can participate and people who have been around for a while can get "refreshers." We offer math workshops on a quarterly basis. Finally, make sure your top person buys in.

How would your advice be different for science centers than for children's museums?

Keith: We're dealing with younger children and their parents in children's museums. In science centers, you're dealing with a somewhat older group, so the level of math will be different. There would also be more ways of integrating science into math and more opportunities to find the math in the science. It also seems that as you develop new exhibit components, you could work on making them more interactive at the same time that you're making them more mathematical.

Maybe people won't be as shocked about doing math in a science center as in a children's museum. There should be a fairly natural movement to add math into the work of science centers.

Principle 2: Do math together, as part of a community

Take one or two of the shapes from Crazy Cakes and ask a friend or family member to work with you on dividing them into halves. You may be surprised at how the experience is different when it is done collaboratively. In school and in informal settings, doing math together often brings up more possibilities, ideas, and solutions than when the work is done in isolation. For example, in Crazy Cakes, people see the shapes differently and often come up with ways of solving the problem *together* that they couldn't have done separately. Doing math together means doing more productive math. Most mathematicians, and an increasing number of teachers and students, collaborate on problem-solving so they can find different ways of generating solutions. Of course, most visitors to science centers come in pairs or groups (rather than individually), so designing experiences for collaborative mathematics is a practical goal as well as an educationally sound one.

Doing math with your colleagues is vital because it helps you establish common ground when planning new mathematical programs and activities, or when trying to highlight the math in an existing exhibit. You need a shared understanding of what it means to do inquiry in math. How can you establish this shared understanding? Find a way of mounting a common math experience for a whole department in your center, or better yet, have everyone in the entire museum—volunteers and staff alike—participate in this experience. It's important to provide not just one experience but also a way of continuing the work. It's possible to do this, even at a large institution, and the rewards are substantial. Here's an example of how it can be done.

Principle 3: Observe math thinking in action

It's essential to have opportunities to observe how visitors get engaged with math, use objects and exhibits to construct math, and "talk math." If you don't recognize exciting math from the visitor's perspective, there's no way to set up experiences and exhibits that foster it. To observe mathematical talking, constructing, and thinking, you first need to identify a few places in your center where it's likely to occur. If you have an exhibit or program that's explicitly mathematical, that's a good place to start, but there are many other places where the math is just below the surface of science activities. Here are some likely places:

✓ Look for exhibits or activities where size is important. Visitors are often interested in the biggest or most, and there can be a fair amount of mathematical talk involving comparing or questioning sizes. For example, we've heard visitors ask, "How much would fit in the mouth of the biggest alligator in this place?" and "What's the biggest bubble you can make?" These questions are cues that mathematics is starting to happen.

✓ Find places that invite comparisons of speed or distance. For example, there may be opportunities to figure out which course for boats is the fastest, how long a penny takes to fall into the gravity well when dropped from different angles, or what circumstances make a Ping-Pong ball stay in the air the longest. All of these elicit mathematical talk, such as, "Let's see which if our balls stays in the air the longest."

✓ Any place with live animals invites mathematical talk about the habits and care of these animals. For example, colonies of leaf-eater ants often elicit talk about how many times the ants travel back and forth, how much they collect in a day, how much bigger the queen is than the workers, etc. With larger animals, there are often questions involving comparison, for example, "How much does this animal eat or sleep compared to me?"

✓ Look for places where visitors can collect data about themselves, such as exhibits on the human body. Visitors who collect data on reaction time, breathing rates, and pulse rates are prone to engage in mathematical talk (e.g., "How much slower is my pulse than that of the average person?" or "How much faster is my reaction time?").

✓ Find places where the concept of time matters and it's challenging to understand. For example, exhibits or programs involving evolution, paleontology, or the nature of the solar system frequently involve thinking about events that occurred in the distant past. Young visitors often wonder, "Did the dinosaurs (or mammoths) exist when grandpa was a child? Or was it right before he was born?" Again, these questions may signal that mathematics is beginning to happen.

Once you've found a place where mathematical thinking and talking are likely, the next step is to park yourself in this place and observe visitors in action. Start by observing pairs of people (rather than larger groups) having conversations as they engage with the activity or exhibit.

What should you look for? First, do visitor observations in the way you have already been doing them. There are several good resources about observing visitors as they think and learn together

(Borun, 1998; Leinhardt and Knutson, 2004; Leinhardt, Crowley, and Knutson, 2002). Second, think broadly about math. If you take the narrow view and look for people doing number operations, you're unlikely to find much that's of interest. Listen for questions visitors are raising that involve comparison, estimation, or prediction, like the ones listed above. As soon as you hear an interesting question that involves mathematical thinking, write down as much as you can about the interchange. If you have a video camera and can get permission of visitors to be recorded, you'll be able to document more thoroughly what happens. In any case, don't try to analyze the conversation on the spot—wait until it's over and you have a chance to reflect. Your job while the discussion or activity is happening is to record everything you can about it that may be of interest.

Once you have a few records of these events, get together with colleagues (preferably colleagues who have also collected observations) and talk about what you saw. Discuss specific parts of your observations, rather than trying to generalize. It helps if everyone in the group has a copy of each other's notes or if you can view pieces of the videotapes together. Here are some questions to get you started:

✓ What did the visitor say that indicates she or he was thinking about a mathematical idea? What was this idea?

✓ How did she or he pursue that idea? For example, did she examine an object more carefully, test a mathematical question by doing an activity (such as constructing a bridge), or talk through her idea with someone else?

✓ What evidence is there that she or he was engaged in the mathematics? For example, did he do several trials of the arch-building activity and make progress in terms of placing the keystone?

✓ How did the visitor use the "object" or activity to do mathematics?

✓ What evidence was there that inquiry was taking place?

As you examine visitor behavior, look for mathematical strengths whenever and however they occur. The challenge is to examine these conversations—which don't sound like school math—and see what is beneath them. Emotional cues, such as eagerness to try an activity or delight in a phenomenon, can also be cues that mathematical learning is occurring.

When you look for math talk and the mathematical use of

exhibits, remember to keep an open mind about who does mathematics. It is just as likely to be the four-year-old girl as the older man with the calculator in his pocket. Math happens in science centers among groups of teens, visitors from the senior center, preschool children, and family groups. Needless to say, it happens among people from all backgrounds, social classes, and races.

Principle 4: Uncover the ways that scientists and other staff at your institution use math in their work

Paleontologists, ornithologists, ecologists, physicists—the whole range of scientists who may be affiliated with your center—are likely to use math and use it a lot. One way you can build the capacity to do math in your science center is to discover the uses of math in the science that you're highlighting. All scientists collect data by observing or measuring, and they use these data to guide their experiments and observations. For example, at the Marine Science Institute in Port Aransas, Texas, marine biologists measure the size of many individual fish in order to track population growth and determine how it changes over time. They can track the age mix of fish by measuring their length, because length correlates with age. Measurements of these slippery specimens must be done carefully in order to get an accurate idea of the characteristics of the fish population at a given point in time.

How does knowing about the ways in which scientists measure fish inform the development of mathematically enriched programs and exhibits? Often a powerful visitor experience in math is one that matches the math in a real scientific experience. As mentioned in chapter 1, most Americans, especially children, are woefully weak at measurement and need considerable practice to improve their skills. Marine biologists take measurements frequently, and these must be reasonably accurate. Here is a case for the mathematics of measurement serving a real and important scientific purpose. Once this connection between mathematics and marine biology was made at the Marine Science Institute, the staff was able to involve visitors in the scientific work of measuring fish and, at the same time, help them beef up their measurement skills. Acquiring knowledge about the mathematics underlying the work of on-site scientists can be a powerful platform for building new visitor programs.

In our experience, the best way to find out how a scientist uses math is to ask. You will have to ask good questions and eventually do considerable translation between the work the scientist is describing and the mathematics that may be accessible and interesting to visitors. But the first step is to learn more about the math that is exciting and necessary to scientists.

Finding out how scientists use math is a useful first step in just about any design work you are contemplating. At the Science Museum of Minnesota, Maija Sedzielarz interviewed a number of scientists while in the planning stages of her work designing math kits for elementary students. She noted that, it was particularly effective to find and work with one scientist on an ongoing basis.

A good way to begin to incorporate more math into your work with visitors is to find out what is happening, mathematically speaking, on the scientific side of your institution. In doing this, you will also develop a broader definition of mathematics and how it is used.

Conclusion

The most important ingredient in getting started with math is the willingness to look more closely at the math around you and to examine how people (including yourself) engage in this math. It isn't necessary to limit yourself to professional situations when doing this work; watch how people use math in everyday situations as well. Notice how people actually use math in activities ranging from working out to remodeling apartments to planning trips around the neighborhood. Especially notice that few people do "everyday math" the way they were taught to do it in school (and, unfortunately, many feel guilty about coming up with more workable strategies of their own). Observe parents having mathematical conversations with their children. Grocery stores and drug stores are a fertile ground for these discussions. As will be seen in chapter 6, everyday math is a good foundation for building exhibits and programs. But the first step is uncovering this math for yourself.

At the start of this chapter, we presented a problem involving halves and noted that there were multiple solutions to finding halves. Here are some of the solutions that we've collected. They aren't the only ones; believe it or not, there is an infinite number of solutions to each of these problems—and there is also an infinite number of incorrect solutions!

Voice from the Field

Working with Scientists to Uncover Mathematics
Maija Sedzielarz, Program Developer, Science Museum of Minnesota

This interview was conducted in July 2005 by Jan Mokros.

Why did you decide to talk with scientists about the work you were doing in math?
Maija: We thought it would connect the real work of scientists with activities that we were designing for kids. We had a specific question: What kinds of math are you doing that could be adapted for a fifth-grade audience? At first, we interviewed every scientist who works for the museum, about eight people. We talked to the head of collections and to a paleontologist, an archeologist, and a hydrologist doing ecological research at our research station on the St. Croix River. We asked them to tell us about how they used math in their work. And they had to do some translation!

We were looking for a good story—something that kids would understand—and we found a few different ones. We felt that we needed to try out these stories. An ecologist on staff was looking at diatoms, and this seemed to relate to the question of scale: What does something look like when it's magnified ten times or a hundred times? It turned out that the diatom piece wasn't possible to do, so we were stuck with just the magnification piece, and this wasn't as interesting to kids.

How did you finally choose a story?
Maija: The paleontologist we interviewed had a great sense of how to talk to kids. Her story was more accessible, maybe because she was able to link it back to what kids would be doing with math. In fact, when Kristy Curry Rogers (the paleontologist) was a child, she used to have arguments with her father about whether you needed to study math in order to study dinosaurs. Kristy's father insisted that you did, but Kristy herself thought that you didn't. It turns out that she was wrong! So she

knows firsthand that math is important.

Kristy's work focuses on studying dinosaur growth and how this varied from year to year, depending on environmental conditions like the availability of food. To do this, paleontologists look at growth rings on weight-bearing bones like femurs. We thought there was a story here and that kids would be interested in connecting dinosaur growth with their own growth.

How does the paleontologist's work connect to what kids are doing in the activity?

Maija: Kristy worked with us to make a graph that shows the growth of a particular dinosaur. Kids can look at the graph and figure out how old the dinosaur was when it died and how much it grew each year. So making and reading graphs is a big part of the work Kristy does and also a part of what we want kids to do. The other piece that we want kids to do—and again a part of what Kristy does—is measure the length of femurs. We had the kids measure the length of their own femurs and the length of femurs of people in their families. We wanted the kids to know that accuracy and consistency matter in science, and I think they got that message.

What was the process of working with Kristy like? What specifically did you learn from her and how did it change your connection with math?

Maija: We interviewed her and read a couple of the articles she published. Kristy was accessible and answered a lot of our questions, especially about how to measure the femurs. It's not like measuring a simple object—there are lots of decisions to be made about where and how to measure. I hadn't realized it was so complicated. Kristy's work is cutting-edge research, and it's important to connect math with this new work in paleontology. I recently read a Scientific American article about how dinosaurs grew so large and how the bones needed to support such a huge amount of weight. I felt like I could connect the math to the science in a way that I hadn't before.

What advice would you give to people at science centers about working with scientists to find out about the math that they do?

Maija: You need to think about a couple of things simultaneously: What math connections can you make with the math that scientists are doing? And what kind of math do we ultimately want in our exhibits or programs? A lot of back-and-forth needs to go on so that it doesn't feel contrived. Scientists need math to do their work, but it isn't something that's imposed. It's an organic part of their work. You need to figure out how to translate that to visitors.

I also found that scientists are eager to talk and very gracious with their time. We couldn't have designed our math activities without first finding out how Kristy uses math. It was a critical step.[2]

[2] For more information about this project, see www.smm.org/mathpacks/.

Crazy Cakes for Two

Solutions for A.

Solutions for B.

Solutions for C.

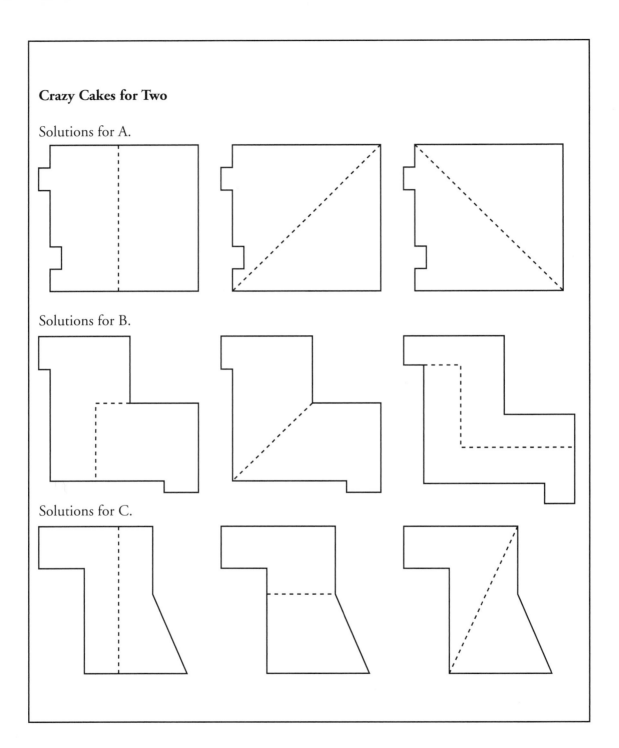

References

Borun, M. et al. 1998. *Family Learning in Museums: The PISEC Perspective.* Washington, DC: Association of Science-Technology Centers.

Gardner, H. 1999. *Intelligence Reframed: Multiple Intelligences for the 21st Century.* New York: Basic Books.

Humphrey, T. and J. Gutwill. 2005. *Fostering Active Prolonged Engagement: The Art of Creating APE Exhibits.* San Francisco, CA: The Exploratorium.

Leinhardt, G. and K. Knutson. 2004. *Listening in on a Museum Conversation.* Walnut Creek, CA: Altamira Press.

Leinhardt, G., K. Crowley, and K. Knutson. Eds. 2002. *Learning Conversations in Museums.* Mahwah, NJ: Lawrence Earlbaum.

Tierney, C., M. Ogonowski, A. Rubin, and S. J. Russell. 1998. Different shapes, equal pieces. *Investigations in Number, Data, and Space.* Menlo Park, CA: Dale Seymour Publications.

chapter 3

The Quest for Mathematical Equity

Jan Mokros and Ricardo Nemirovsky

In the 1980's, the federal government issued a report about the tragic failure of math and science education. The most notable and disturbing line is this: "If an unfriendly foreign power had attempted to impose on America the mediocre educational performance that exists today, we might well have viewed it as an act of war. As it stands, we have allowed this to happen to ourselves" (National Committee on Excellence in Education, 1983, p. 1). The report galvanized attention to the fact that U.S. schools were working hard to produce an elite group of students skilled at math and science, while at the same time filtering out most students from these disciplines. It is still the case that at each successive grade band, more students are filtered out of math so that not many remain by the end of college or graduate school. In other words, math serves the same purpose that Latin used to serve— to weed out students who are not, for whatever reason, doing intense academic work.

The committee posed another way of thinking about mathematics: Instead of thinking about math as a filter, it could be thought about as a pipeline that could bring as many people as possible into as many areas of math as possible. The message was clearly about keeping people in math, and for good reason. It was pointed out twenty years ago and is even truer today that the United States has been losing ground with respect to building a strong scientific and mathematical pool of talent. According to myriad recent reports, the United States is being surpassed by several other countries in preparing students for careers in math, science, engineering, and technology (Friedman, 2005; Business Roundtable, 2005). A group of major business leaders has concluded, "Together, we must ensure that U.S. students and workers have the grounding in math and science that they need to succeed and that mathematicians, scientists, and engineers do not become an endangered species in the United States" (Business Roundtable, 2005, p. 8). This report stresses the need for a special effort in recruiting and retaining currently unrepresented groups in mathematical and scientific occupations.

Looked at from an individual's perspective, there are many benefits to being fluent in math. People who take more mathematics courses earn substantially more money than people who do not, and have more career opportunities. Murnane and Levy (1996) point out that at a minimum, people need the ability to do math competently at a ninth-grade level (through algebra) in order to get well-paying jobs. Even assuming that money and careers are not a major goal for everyone, it is important to keep people involved in math for other reasons. Mathematically skilled people are better able to estimate costs when they are shopping, determine how much an item on sale should cost, interpret medical findings, understand how to do their taxes, interpret political polls, and make consumer decisions.

People who have been filtered out of mathematics are not representative of the U.S. population as a whole. Not everyone has equal access to math, and students who drop out are disproportionately people from low-income, African-American, and Latino groups (Trentacosta and Kenney, 1997). Until recently, women were also less likely to pursue math, though by 2000, 47% of all math majors were women, and women and men were achieving comparable grades in college math courses (Spelke, 2005).

It's important to dispel the myth that any demographic group has less innate mathematical talent than another. The most that can be concluded from an immense amount of research is that some groups of people perform better on some tests of some mathematical skills under some conditions—and that these differences are highly related to a group's resources and income. For example, there is an exceptionally high relationship between family income and U.S. high-school students' scores on the math (and verbal) SATs (Fairtest, 2003). Income is by far the best predictor of how well a student will perform on SATs. Wealthier families have access to a variety of educational resources, including well-funded schools and tutoring, that are simply not accessible to families with less income. Jonathan Kozol talks about this situation being maintained by "savage inequalities" in local education funding (Kozol, 1991). He has recently written about this country's schooling system that generates and maintains, in essence, racially apartheid schools (Kozol, 2005). There isn't a level playing field when it comes to involvement in math.

School-based educators have been approaching the problems of inequity from various directions. Some programs, such as the Algebra Project, are specifically targeting underrepresented groups (in this case, African-American middle-school children) with high-quality, rigorous mathematical curricula. By working intensively with a group that hasn't achieved as much as they have a right to achieve, greater fairness could be gained. Others have designed special courses and programs that focus on involving talented students from underrepresented groups (AERA, 2004). The government initiative No Child Left Behind, from the early 2000's, has approached these inequities by mandating a higher level of teacher preparedness in math and regular testing to identify achievement gaps among various subgroups.

A promising approach to dealing with inequity in mathematics in schools, which offers many lessons for science centers, is that of Eric (Rico) Gutstein, who advocates combining learning about math with learning about social injustices (Gutstein, 2006). Gutstein's work with middle-school Latino children in urban Chicago builds upon real-life projects that are mathematically rigorous and politically meaningful to children. For example, projects have dealt with racial profiling, figuring out how many students could receive full four-year college scholarships for the cost of one B-2 bomber, and the consequences of "driving while black or brown." All of these questions are of interest to older students and adults who have been oppressed. It is useful to consider how science centers could build mathematical experiences from a foundation of truly critical social justice issues.

The Need for Mathematical Equity in Science Centers

Science centers could be key players in addressing inequities with respect to math achievement. Achieving "sustainable diversity" (Jolly, 2002) is a broad and critical goal for centers, one that goes far beyond the boundaries of math or any other discipline. The bylaws of the Association of Science-Technology Centers support bringing those who have been traditionally underrepresented in science and math into the enterprise of lifelong learning in these disciplines. ASTC has developed a diversity toolkit to assist science centers in addressing the needs of increasingly diverse audiences (ASTC, 2001). The focus of much of the organizational diversity work has been on bringing more diverse audiences to centers, working more effectively with a range of audiences through community partnerships, and on staff recruitment and professional development. In other words, science centers have done a great deal of work in bringing underserved audiences into their centers and in reaching out to these audiences through community programs. But how do institutional missions to achieve greater diversity intersect with centers' goals to deepen mathematical programming? How can science centers achieve equity when it comes to their work in math?

First, it's important to consider what mathematical equity might mean within science centers. One way of achieving equity might be to "treat everyone exactly the same." This would mean providing the same accessible exhibits for everyone, with the same types of mediation, the same availability of explainers, and the same easy-to-read signage. These are important starting places, but they fail to address the need for different approaches to different individuals and groups. A prominent equity educator, Walter Secada (1999–2001), believes it better to focus on fairness, rather than on treating everyone the same way. In his view, treating people fairly means taking into account the differences among them and explicitly addressing these differences. This may include attention to different mathematical and cultural backgrounds and experiences, learning styles, and needs for time and help.

Science centers have several advantages in the quest for mathematical equity. First, they are built around the premise that individuals need to be engaged in science that is based on their own interests. Second, they are respectful of visitors' expertise and build upon this in creating exhibits and programs. Third, they create connections between the different "languages" of their visitors, both in the cultural sense and in terms of connecting everyday language with more formal scientific language. Turning these advantages into tools for producing greater mathematical equity is the focus of the following discussion and examples.

Equity through relevance

Science centers have a critical tool to address equity—the ability to engage visitors in relevant math that is seen in a variety of intriguing contexts. This tool is one that formal educators are trying to use more often and more effectively. In one of her President's Messages to the National Council of Teachers of Mathematics, Cathy Seeley states, "Student engagement is perhaps our most important tool in our battle for equity" (Seeley,

2004, p. 3). She goes on to talk about the importance of providing tasks that a broad range of students will find relevant. Here, we illustrate in detail how one school program, built upon an exhibit that recreates elements f a Hmong house, pulls children into math by engaging them with cultural artifacts that are a significant part of their lives.

This case is based on a school program at the Science Museum of Minnesota that focuses on materials developed for a math project. The project developed curricular materials to support math activities during field trips to the museum by fifth-grade students. It included classroom activities for before and after the field trips. One important goal of the program was to reach urban elementary students who needed additional academic support. Some of the school visits were videotaped and analyzed, as was the ten-minute episode presented here, which took place in summer of 2004.

At the beginning of the episode, the fifth-grade students, including several from Hmong immigrant families, are sitting on the floor reading a student sheet about the mathematics of tessellations, with the assistance of a teacher. The student sheet explains tessellations as "a pattern of closed shapes that completely covers a surface," and it shows samples of tessellations from clothes used in different cultures. The activity consists of looking for tessellation patterns in the museum's Collections gallery, drawing one of them, and discriminating its geometrical components.

The students in this episode are exploring the Hmong house exhibit, where they have found tessellation patterns of triangles that form the border for a Hmong story quilt. Once inside the house, one of the Hmong girls, Elizabeth, stops in front of a reproduction of a traditional Hmong altar. Elizabeth sits on a bench in front of the altar and calls to some of the other students to join her. While Elizabeth calls them, she bounces slightly on the bench. The teacher asks Elizabeth, "So, do you see some tessellations here?" Elizabeth responds, "Yeah, right here," pointing at a strip of paper cut into a pattern. Elizabeth stays on the bench, drawing the pattern on the student sheet.

Shortly thereafter, another student walks up and asks Elizabeth about the altar. Elizabeth responds, "It's a Hmong thing that Hmong people…they, uh…do this; they light that (pointing to something in the altar), and then they, uh, say a little prayer (bouncing on the bench) so the ghosts don't come back to them and haunt them forever."

The girl looks at the altar a moment longer and then

walks away. The teacher sits down next to Elizabeth.

Teacher: So, have you been to a shaman's house, where you…
Elizabeth: That's my grandpa.
Teacher: He's a shaman? (Elizabeth nods.) Awesome.
Elizabeth: Um, my grandma's a shaman, too. Um…like if you…like, say, they're not… people are not supposed to [do something], and they say a little prayer (slight bounce), and they say that…like, help them (bouncing). Like, they light that (points to object in altar). They. . . um, they put all this right here (gestures horizontally on the altar, over the upper shelf).
Teacher: So it looks a lot like this (points at the altar).
Elizabeth: They go (she bounces)…they create, so souls…so ghosts…don't haunt them.
Teacher: So people will come to them.

Tessellations around a Hmong story quilt.

MATH MOMENTUM IN SCIENCE CENTERS

Elizabeth: So ghosts don't haunt them forever. Like if they did something.

Then a Hmong boy, Robert, walks over and puts his clipboard on the floor. Elizabeth gets up and the boy sits down in her place.

Robert: So, they put something over the head, and they do this (he bounces like Elizabeth did, only higher).

Robert gets up and Elizabeth sits down again. She starts bouncing while the teacher and the boy are talking. Elizabeth says, "My grandpa does that," and she starts drawing on her clipboard.

What does this episode tell us? Within the Hmong house, Elizabeth and Robert became authoritative sources for the meaning of the altar; they were visitors who could animate the altar to explain it to the teacher and other children ("It's a Hmong thing," said Elizabeth). This interaction suggests a challenge that goes to the core of equity: designing math programs and exhibits so that visitors from nonmainstream worldviews can participate as experts and contributors.

Gutstein (2005) talks about the importance of blending classical or traditional mathematical knowledge with community interests. In the Hmong house, Elizabeth and Robert are beginning to learn classical mathematical knowledge—in this case, knowledge of geometry—within the context of deep interest and knowledge about their community. Equity is not only a matter of making programs and exhibits accessible to all, but also involves enabling minority visitors to share their own cultural traditions. In other words, equity opens up a space for all visitors to encounter ways of talking and acting that fall outside of what is considered mainstream culture. The challenge for science centers is to actively combine community knowledge with classical knowledge to create a "pedagogy of access" (Morrell, 2005) that enables more people to enter mathematics.

Another important message in this episode involves the question "How does one make a highly engaging exhibit mathematical?" A good starting point is by carefully examining an exhibit (see chapter 5) to find the mathematical opportunities. In the Hmong house, the mathematics of tessellation in the story-quilt patterns was made explicit through the use of student materials and worksheets. The math was initially hidden within this important exhibit about Hmong culture. This suggests that when thinking about equity, one place to begin is with existing exhibits that have strong cultural components. The next step is to identify places where mathematics might be lurking in these exhibits and to incorporate meaningful mathematics into them. As seen in this episode, integrating more math need not be an expensive proposition and doesn't have to involve building more exhibit components.

Finally, this episode raises questions about what makes an exhibit not only engaging, but also interactive. In the science-center world, it is common to refer to an exhibit as interactive if it has buttons or other mechanical-electronic means to respond to the user's actions. In the Hmong house, we encounter a reproduction of an altar that does not have any buttons or ways to mechanically elicit responses. And yet, for Elizabeth and Robert, the exhibit was a call to action. They bounced on the bench in a particular fashion that recreates some of the rituals performed facing the altar; they gestured to the objects positioned on the shelf and talked about their function; and they acted the presence of their ancestors striving to alleviate those who had done something unjustifiable.

Interactiveness—the genuine mutual animation of visitor and exhibit—is not a matter of buttons, but of a resonance between the exhibit and the life background of the visitor. An exhibit is not interactive *per se* but *for someone*, and it relies upon the awakening of significant experiences leading him or her to animate it. For Elizabeth, the idea of tessellations came to be interwoven in the enactment of traditional prayers; she transitioned back and forth between drawing the pattern visible on the paper strip and bouncing on the bench to imitate the work of a shaman.

Equity through capitalizing on visitor expertise

Connected to the issue of engaging visitors is the issue of capitalizing on their expertise. All visitors, no matter how young or old, or how poorly or well they have done in school math, have mathematical expertise. Whether they are able to use this expertise in science centers depends on how exhibits and programs are designed. For example, visitors may have deep knowledge about the mathematics of measurement as used in gardening, remodeling, or cooking. They may use math in occupations as diverse as plumbing and nanotechnology; and they may use visual-spatial skills in all sorts of artistic pursuits. By building on visitors' expertise, they gain access to math in ways that are nonthreatening. For example, the Math in the Garden program at Lawrence Hall of Science takes visitors of all ages directly into the garden and engages them in familiar planting and monitoring tasks that have a mathematical bent. Young children are given ten popsicle sticks and choose ten plants to examine. Each plant that is not "eaten" in any way gets a blue stick, while plants that have been nibbled upon get a red stick. Because young children are working on establishing what two numbers can be combined to total ten, this task is an excellent way of melding gardening with developmentally appropriate mathematical reasoning.

Visitors bring all kinds of math-related knowledge with them on their visits to science centers. Finding and mobilizing cultural "funds of knowledge" with respect to mathematics is a critical task of science centers (Moll and Gonzalez, 2004). According to the Funds of Knowledge approach, the first step is to investigate community knowledge, focusing on the strengths that people bring to an educational situation. Moll offers a list of questions for educators as a self-assessment to guide this process. These questions are adapted here to address science-center audiences, specifically with respect to mathematics.

Building on Visitors' Funds of Knowledge

1. How well does our center link visitors' math experiences to families and communities?

2. Do we provide ongoing parent education and training so parents can help their children with math?

3. Have all our staff had training to help them use visitors' families, languages, and cultures as a foundation for learning?

4. How do our staff members tap into visitors' funds of knowledge?

5. In what ways do we affirm visitors' home languages, while linking them to Standard English?

6. Do our staff members know how to use visitors' informal languages as a tool for developing math literacy?

7. How well does our center tailor its exhibits and programs to the particular needs, interests, and learning styles of individual visitors?

8. In what ways do we encourage and teach to the many intelligences and learning styles of visitors?

9. How does our center encourage visitors to articulate their dreams and aspirations and link them to math learning?

Science centers know how important it is to build upon visitors' strengths and interests, and not to prejudge what visitors can do. Making use of what visitors bring to the situation, as Moll suggests, means truly knowing visitors' backgrounds and skills. The best way of understanding visitors' expertise is to spend a substantial amount of time talking with them informally and through more formal visitor studies. Another rich method of understanding what visitors bring to a situation is to interview them in their homes. When Moll undertook such a home study, he discovered a great deal of "hidden family knowledge" in the local Latino barrio, knowledge that the schools did not know about and could therefore not use to teach academic skills (NCREL, 1994). As Hein points out, "It is a good rule of thumb to assume that you know less about your visitors than you think you do" (Hein, 1998, p. 164). This is especially true in the subject area of mathematics.

Hein also suggests that working with visitors' expertise and passions can be informed by considering extreme situations or by thinking about visitors with special attributes. The question "What kinds of unrecognized strengths might visitors have?" is especially interesting when considering unusual strengths or unusual combinations of attributes. For example, the award-winning 2005 documentary *Murderball* rivets viewer attention on the unanticipated, seemingly incongruous strengths of a group of quadriplegic athletes. These men are aggressive, fierce competitors involved in elite international competition in the sport of full-contact rugby. They have unexpected skills that most able-bodied people would never have imagined. The documentary makes use of interviews and close observation (and wonderful photography) to reveal these skills.

To continue with this context and connect it with mathematics, a thought experiment is useful: How might knowing about this group's unusual athletic prowess guide our thinking about materials, programs, and exhibits that would draw them into math? Obviously, this is a quite different question from the one that we might first think about, namely, How do we make a math exhibit accessible to people in wheelchairs? Physical accessibility is essential, but it is not the only starting point. Building an exhibit or program around the talents of wheelchair athletes is a quite different dimension of equity. Wheelchair athletes, whether marathoners, rugby players, or skiers, probably bring a great deal of expertise at the kinesthetic level about the mathematics of torque, velocity, force, and angles. They understand many aspects of movement and are connecting the physics and mathematics of motion as they engage in their sports.

The *Murderball* example is admittedly an extreme one, involving a very small audience. The point is not to rebuild exhibits to meet the needs of physically handicapped individuals (though this is an important goal), but rather to uncover the unrecognized mathematical strengths of any group of visitors to your institution. You may find that middle-school children have developed extraordinary pattern-finding skills as a result of their participation in the current dance crazes, and these skills could be the basis for a mathematical activity. Or, you may find that the sewing skills of a talented group of Latina women offer a way into any mathematics that involves spatial visualization.

Beginning the work of designing mathematical experiences by examining visitors' expertise represents a major shift in perspectives. As Martin and Toon point out, "a kind of Copernican revolution has taken place in the type of enculturation museums do. The locus of meaning-making has shifted from being centered on the museum's body of knowledge to the museum's understanding of its visitor" (Martin and Toon, 2005, p. 409). The power of *Murderball* as a documentary is that it begins with an understanding of the full identities and talents of its characters.

Equity through multiple languages

There is a strong and growing need to reach English language learners with mathematics (National Science Board, 2006). Many of the exhibits and programs designed by science centers are translated into the major languages spoken by visitors. In the science centers represented in this book, for example, nearly all have translated signage and materials into Spanish. This is especially true in centers with large Latino populations. For example, the Miami Museum of Science serves large numbers of visitors who speak and read Spanish fluently, but who speak very little English. These visitors need Spanish signage, and because they are very educated, the signage needs to meet their expectations for sophisticated scientific content. In other situations with Spanish-speaking visitors, reading skills may not be as developed, and more simplified signage is called for.

Using the native languages of science-center clients is a good strategy because it gives visitors multiple entry points. This approach also shows respect for language diversity and may help visitors feel more at home in science centers. However, the approach needs to meet specific needs of specific subgroups. Before proceeding with translating materials, it is important to understand the audience.

Voice from the Field

More to It Than Meets the Eye: Translating Materials to Spanish

Carlos Plaza, Exhibit Developer, Miami Museum of Science

Our science center's audiences include a broad range of Spanish-speaking visitors from Cuba, the Dominican Republic, Argentina, Guatemala, and Mexico, to name a few. Many of them are very fluent in written Spanish. A problem in translation from English to Spanish is that sometimes there is more than one word for the same object in common use among various Spanish-speaking populations. This means that translating even a single sentence may require using several reference books and personal contacts. In our work with the mathematics of measurement, we focus on caimans, which in English means an alligator-like reptile. But the translation is complex because the word is used quite specifically in the Amazon basin, where the animals are ubiquitous. The word caimans is also used more generally in kids' songs. There are several Spanish dictionaries that translators must use to figure out the possible translations, before using judgment to decide which is the best match for a particular audience. Many Spanish-speaking people in the United States refer to an alligator as a caiman. In fact, my Larousse Spanish Dictionary erroneously defines the alligator as a synonym of caiman. The entry reads: "ALIGÁTOR. s.m. (ingl. Alligator). Caimán." In the very same dictionary there is an illustration for the entry "CROCODILIOS" (crocodilians) depicting a gavial, a crocodile, an alligator, and a caiman. The first erroneous entry may reflect the common usage—if it looks like an alligator, it's a caiman and vice versa—although the illustration presents a scientifically accurate description!

Another example of there being more than one word for the same thing is "plumbing"; I've actually had to use that word in an exhibit. Here are some of the different ways plumbing is translated in Spanish-speaking countries:

fontanería = Spain
plomería = Argentina, Cuba, and Mexico
gastifería = Chile, Ecuador, and Peru

Multiple meanings of the same word can sometimes prove embarrassing to exhibit builders. The word bagre means "catfish" in many Spanish-speaking communities, and one can order it in a restaurant. In other communities, bagre means "ugly woman" (colloquial), and in those places, it can be insulting to order bagre in a restaurant. It is appropriate instead, to use the word pescado, which is used to refer to catfish, as well as any other fish. (You can also say pez gato, which literally means "cat-fish," and is used in many countries that don't use bagre to refer to the fish).

All of these subtleties need to be taken into account from the beginning of a project, not crammed into the last few weeks of preparation. The most effective way to do this is to be aware of the requirements of translation as you write English copy. Then, as soon as possible, the bilingual translator should begin to work on the text. Often, as the translator works to convey the meaning of the English text, it becomes clear that the language could be simplified and clarified. And clarity is the goal in any language! A broader view of language can expand the strategies we consider for approaching translation issues. Marta Civil of the University of Arizona points out that there are at least four languages at play in formal and informal education. Some are more important for academic success than others, while some might be more effective in communicating mathematical and scientific ideas (Civil and Planas, 2004). It is important to consider all of the following types of languages:

✓ Mathematical language, including both words and symbols

✓ Physical or body language, such as that used by children in the Hmong house exhibit

✓ Everyday language —the language of friends, families, and the local community

✓ Academic language—the language found on tests and required in English essays, a language that not all people have, regardless of their first language

Visitors of different cultural and language backgrounds may actually share more physical or body language than do two people who grew up speaking the same everyday language. Written English academic language may be foreign to some English-speaking students, but very clear and engaging to certain Spanish speakers. Mathematical language and symbols, although often thought of as universal, are far from universal to visitors who dislike math, regardless of their cultural backgrounds. All of these languages need to be considered when making design decisions.

If Howard Gardner's ideas about multiple intelligences (Gardner, 1999) are adapted to Civil's on multiple languages (Civil and Planas, 2004), the resultant theory suggests that different visitors have strengths in different languages. This includes the physical language of movement. The aim in designing programs and exhibits is to pull as many people into mathematics as possible by using multiple entry points. Moreover, if we aim to advance visitors' understanding of mathematics, it is important to make connections among the languages that are more familiar (i.e., everyday language and physical language) and those that are less familiar (i.e., academic language and mathematical language). The issues here are complex, and useful approaches will require new conceptual frameworks for communication and learning.

Conclusions

This chapter outlines three starting points for science centers as they work on creating mathematical equity. One such is to engage visitors by building upon their interests. A second, related starting point is to make use of visitors' own mathematical expertise, including their community's knowledge, and to foster visitors' understanding that they are capable of constructing mathematics. Mathematics, in this view, is a part of each person's interests and cultural background. A third place to begin is by using a variety of languages—including non-English languages, everyday language, physical language, mathematical language, and academic language—and to connect these languages. This gives visitors multiple ways of entering mathematics. All three of these starting places lead to greater engagement in math. Of course, engagement is only the beginning. Providing high-quality mathematical experiences—experiences in which mathematics is seen as beautiful, useful, politically empowering, challenging, and scientifically necessary—is our overall goal. To achieve this goal, visitors' "interaction with the contents of the museum must allow them to connect what they see, do, and feel with what they already know, understand, and acknowledge. The new must be able to be incorporated with the old" (Hein, 1998, p. 153).

References

Association of Science-Technology Centers. 2001. *Equity and Diversity Toolkit*. Washington, DC: ASTC.

Business Roundtable. 2005. Tapping America's potential: The education for innovation initiative. Downloaded from www.businessroundtable.org, April 18, 2006.

Civil, M. and N. Planas. 2004. Participation in the mathematics classroom: Does every student have a voice? *For the Learning of Mathematics* 24(1):7–12.

Fairtest. 2003. SAT race, gender gaps increase. *Fairtest Examiner*. Cambridge, MA.

Friedman, T. 2005. *The World Is Flat: A Brief History of the 21st Century*. New York, NY: Farrar, Straus, and Giroux.

Gardner, H. 1999. *Intelligence Reframed: Multiple Intelligences for the 21st Century*. New York: Basic Books.

Gutstein, E. 2006. *Reading and Writing the World with Mathematics. Toward a Pedagogy for Social Justice*. New York, NY: Routledge.

Hein, G. 1998. *Learning in the Musum*. New York, NY: Routledge.

Jolly, E. 2002. *Eric Jolly on Sustainable Diversity in Science Centers*. VHS video and 65-page guide. Washington, D.C.: Association of Science-Technology Centers.

Kozol, J. 1991. *Savage Inequalities: Children in America's Schools*. New York, NY: Crown Publishers.

Kozol, J. 2005. *The Shame of the Nation: The Restoration of Apartheid Schooling in America*. New York, NY: Crown Publishers.

Martin, L. and R. Toon. 2005. Narratives in a science center: Interpretation and identity. *Curator: The Museum Journal* 48(4):407–426.

Moll, L. C. and N. Gonzalez. 2004. A funds-of-knowledge approach to multicultural education. *Handbook of Research on Multicultural Education*, 2nd ed. San Francisco, CA: Jossey-Bass:699–715.

Morrell, E. 2005. Doing critical social research with youth. Talk given at DePaul University, Chicago, IL, February 3.

Murnane, R. and F. Levy. *Teaching the New Basic Skills: Principles for Educating Children to Thrive in a Changing Economy*. New York, NY: The Free Press.

National Assessment of Educational Progress. 1999. Long-term trend assessment. Washington, DC: U.S. Department of Education, NAEP.

National Commission on Excellence in Education. 1983. *A Nation at Risk: The Imperative for Educational Reform*. Washington, DC: U.S. Department of Education.

National Science Board. 2006. America's pressing challenge—Building a stronger foundation. Washington, DC: National Science Board Press.

North Central Regional Educational Lab. 1994. Funds of knowledge: A look at Luis Moll's research into hidden family resources. *Cityschools* 1:19–21.

Schliemann, A. et al. 2006. *Bringing Out the Algebraic Character of Arithmetic: From Children's Ideas to Classroom Practice*. Hillsdale, NJ: Lawrence Erlbaum Associates.

Secada, W. G. Ed., series and vol. 1. 1999–2001. *Changing the Faces of Mathematics*. 6 vols. Reston, VA: National Council of Teachers of Mathematics, Inc.

Seeley, C. 2004. President's message. *NCTM Newsletter*, November, p. 3.

Spelke, E. 2005. Sex differences in intrinsic aptitude for mathematics and science? A critical review. *American Psychologist* 60(9):950–958.

Trentacosta, J. and M. J. Kenny. Eds. 1997. *Multicultural and Gender Equity in the Mathematics Classroom. The Gift of Diversity*. Reston, VA: National Council of Teachers of Mathematics, Inc.

CHAPTER 4

Mathematical Challenges in Science Centers

Jan Mokros

What is a mathematical challenge and how can such challenges be incorporated into the work of science centers? As noted in chapter 2, good mathematical challenges in science-center, zoo, and aquarium settings involve visitors and staff in inquiry and engagement. Such challenges are not simply about getting the "right answer," if in fact there is a single right answer. As we have seen, a real mathematical challenge may involve several right answers or several different approaches to a solution.

As you begin to incorporate math into your work, the difficult, yet essential task is to change the way you think about a mathematical problem. Instead of basing your work on the word problems or other problems you encountered in school, think about challenges the way that mathematicians do. Most mathematical challenges demand creativity and coming up with an effective, yet clear and generalizable path toward solution. Unlike school problems, real math challenges are not based on how quickly one can calculate or on how many formulas one has memorized.

To start thinking about math challenges for visitors, consider a familiar activity that visitors encounter in aquariums and zoos: determining how wide your own "wingspan" is and comparing it with the wingspans of various birds. The idea is to show visitors that even birds they may consider to be small, such as crows, have impressive wingspans, and that wingspans of birds are proportionately much longer than their body lengths. (In humans, by contrast, wingspan width and height are about the same.) When visitors engage in a wingspan activity, the math typically involves a quick comparison. Sometimes visitors are simply asked to note whether their wingspans are bigger or smaller than that of different birds; sometimes they are asked to measure and determine how much bigger one of the wingspans is than the other. These are good beginnings to math problems, but they don't go far enough in challenging visitors to engage in significant mathematics. They are familiar problems, maybe *too* familiar. "How much smaller is a crow's wingspan than yours?" is a lot like the word problems found in math textbooks. This is especially the case if one is asked to simply compare two numbers (e.g., your own wingspan and that of a prototypic crow). There is more context and more visual support for the activity in science centers and zoos, but it may not be a compelling problem, let alone a mathematically challenging one.

Now consider the following activity: A zoo education staff member has collected data on the wingspan widths of several crows, including some babies. She has made sticks representing the widths of these wingspans. In making this data set, she has gathered real data from a collection of crows that has been measured and tagged at a research station. She has made a comparable data set consisting of the "wingspans" of several different humans. The crow wingspan sticks are painted a different color than the human wingspan sticks, for easy identification.

There are several mathematical challenges that can be posed with these two data sets. For example, for elementary-school visitors, a challenge is to put the crow wingspan sticks in order and see what conclusions can be drawn. One can do this with the human data set as well, and visitors can be encouraged to address the question, "How does your own armspan width fit into the data set for humans?" Note that to figure out the answers to these questions, a visitor needs to order, or seriate, the data set, an important measurement challenge for young visitors. Somewhat more complex challenges can be posed for older visitors, especially challenges involving comparisons between the two data sets. For example, comparing the widths of crow wingspans with those of humans raises mathematical questions such as these: Are the wingspans of all crows in this data set smaller than the armspans of all humans? How much variation or difference among individuals is there in the data sets? In most cases, data sets are complicated, and there is considerable variation within them. Involving visitors in the mathematical challenge of accurately describing and comparing data sets not only fascinates them, but also immerses them in critical mathematical ways of making sense of data.

Ingredients of Math Challenges in Science Centers

Math challenges should have as many of the following ingredients as possible:

✓ Start with engaging and motivating questions of interest to a wide range of people

✓ Prompt inquiry into questions where the answers are not known in advance and where there is a purpose for finding the answer

✓ Encourage discussing, sharing, and collaborating

✓ Connect to a larger scientific phenomenon

✓ Engage people with substantial mathematics

✓ Integrate a physical activity with a mathematical challenge

Consider the sketch of the crow-human wingspan activity provided here as an example of a mathematical challenge. Although only a brief outline has been presented, the activity can be evaluated in terms of the six ingredients of math challenges in science centers, as follows.

Start with engaging, motivating questions

These wingspan measurement activities may or may not involve questions that motivate visitors to engage in math. A visitor study would be needed to see whether people care about crow characteristics and comparisons between themselves and crows (or other birds). If they care about the comparisons, the math can be built upon this foundation.

Prompt inquiry into questions with unknown answers

Certainly the answers to most of the questions posed here are not known in advance, though scientists who study crows may know some on a general level. It is unlikely that visitors would feel that they are expected to find one correct answer, as there are a great many comparisons that they could choose to explore.

Encourage discussing, sharing, and collaborating

In this activity, one can imagine parents helping young children put the wingspan sticks in order, or middle-school visitors involved in a discussion about which males—crows or humans—have bigger wingspans. Because the data sets overlap, there is certainly room for discussion and sharing different perspectives.

Connect to a larger scientific phenomenon

There are a number of scientists who study the characteristics, morphology, and behavior of crows. The mathematics of examining these data sets could easily be tied to the work done by scientists.

Engage people with substantial mathematics

It is likely that visitors will do mathematics when they engage with this activity; however, a formative visitor study would show the extent and frequency that this is done by different kinds of visitors. Possibly, visitors will engage to a greater extent in the math when the activity is mediated. The mathematical impact, with or without mediation, could also be explored in a visitor study.

Integrate a physical activity with a mathematical challenge

In this example, a physical activity—arranging and rearranging the sticks while comparing them—is a crucial piece.

It is a goal for an activity to involve substantial mathematics: The wingspan activity focuses on the math of measurement and data. As noted earlier, these two areas are underrepresented in school, and they offer a natural connection to scientific phenomena. The *NCTM Standards* (2000), as well as educational researchers, more generally assert, "The study of data and statistics should be firmly anchored in students' inquiry" (Lehrer and Schauble, 2002). Data and measurement go hand in hand, as measurement is an important means of collecting data. Coincidentally, measurement is one of the areas where U.S. students perform most poorly and where they need a great deal of practice that they are not getting at school (Clements and Bright, 2000). Working with data and measurement is a wonderful starting place for building mathematical challenges into both exhibits and visitor programs.

Involving Visitors in Mathematical Challenges

The wingspan activity is not one that we have seen implemented. Instead, it is a thought experiment that shows how a typical science-center activity can be transformed into a significant mathematical challenge. Next, we present actual mathematical challenges, including two that TERC has designed and science centers in the Math Momentum project have implemented in many different settings. Each of these challenges involves familiar scientific content and is bu0ilt upon the ingredients discussed above.

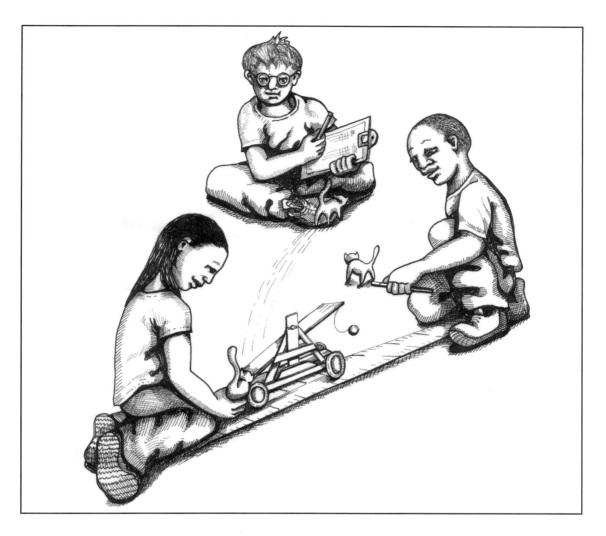

Catapulting

An engaging data and measurement challenge involves teams in setting up a chain reaction of catapults. Small groups each have a Cat-a-pult[1] with several settings that control force and angle of launch of a small rubber cat. They also use a meter stick or a paper tape measure. Groups work on finding a combination of force and angle settings that will result in a predictable distance. Finding this distance is critical, as once each group has taken enough data to establish a reliable setting for its catapult, the members work with other teams to form a chain reaction. Each group uses its data to place its own catapult in relation to the next one in the chain, so that the landing pad of Cat A is the

Children measuring distances of catapult launches.

[1] Cat-a-Pult is produced by HandsOnToys, www.handsontoys.com, Wilmington, Massachusetts, ©2001.

spring-loaded launching pad of Cat B, which in turn lands in a place that launches Cat C, etc.

Success is measured by the number of cats in the chain reaction or by the ability to make a chain of catapults that eventually launches one of the cats into a particular target (such as a small dish of cream!). Errors in data collection and measurement can easily be compounded and prevent a successful chain reaction. To do this work successfully, careful measurement is necessary.

The problem posed above—making a chain reaction of several catapults that consistently works—has all six of the specified ingredients of an effective math challenge. The problem is certainly motivating for visitors ranging in age from upper elementary school through adults. There are many different ways of predicting distances and making the catapults hit their marks. The solution to making them work is not known by anyone in advance, including program staff who are implementing the activity. Although it is conceivable (though difficult) to do the catapult challenge individually, there is a true advantage in tackling it collaboratively and discussing strategies. The physical element makes it more interesting and also means that a trial-and-error approach can be integrated with the underlying mathematical principles. Exploring angle of launch and force is also a way of integrate important mathematical and physics principles.

One other feature of catapulting challenges that makes them appealing in science centers is their flexibility. Catapulting can be done in the context of large groups events, or with younger children, or in the context of a stand-alone exhibit. With large groups, it is easy to envision setting up a catapulting chain-reaction event. A small team works together on the original challenge; then teams set up longer and longer chains. When working with younger children, it is easier to adapt the work so it involves only one step in the chain. For example, Shari Hartshorn at the Science Museum of Minnesota used the catapults with children in primary grades. She asked them to measure the distance the cat would fly and then to set up the catapult so that it would launch as close as possible into the bulls-eye of a target. This was a challenging and engaging activity for younger children, and it gave them a great deal of practice with linear measurement. Another possibility is using the bulls-eye-and-target-practice version of the catapulting activity in the context of an exhibit. The focal point of the exhibit would be the target, and visitors would engage in measuring to determine the location from which they would launch their catapults.

What is the important mathematics that visitors learn while they are engaged with catapults? At the most basic level, visitors get more practice measuring in real situations. Measurement demands knowing where to start, where to finish, how to measure in a direct line, how to "iterate," or move the measuring tool without leaving gaps, and knowing how to read and interpret the results. (For example, what do the tick marks between inches or centimeters stand for?) At a more advanced level, the task involves thinking about variability and measurement error. The catapults don't always travel exactly the same distance, even when conditions are as controlled as possible. How much variability can you expect under the same conditions? Does this differ, depending on the force and angle settings of the catapult? As soon as multiple measurements are involved, visitors are thinking about data and how to examine the reliability of data taken under a certain set of conditions.

The chain-reaction catapulting activity has been one of the most popular parts of staff development workshops implemented by the Math Momentum project. Evaluation data show that participants find the activity particularly effective for these reasons:

✓ It offers an "existence proof" that mathematical challenges can be exciting and involve inquiry.

✓ It is based on a physics phenomenon (e.g., catapulting) that is present in nearly all science centers, so that it is easy to envision conducting similar activities with visitors.

✓ It provides opportunities to talk about mathematics in order to solve a real problem.

Gears

A second example of an effective mathematical challenge in science centers involves gears and ways they can be used to encourage visitors to think more deeply about ratio and proportional reasoning. This challenge emerged from a question posed by mathematics educator and Math Momentum advisor Dr. William Tate: How can science centers create exhibits and activities that help people engage in proportional reasoning? Proportional reasoning is a critical mathematical skill, typically taught and learned during middle school. It is one of the foundations for algebra and higher-level math. The principle is that as one variable changes, a second variable changes in a way that is proportional to the first. At a basic level, to make enough food, you might double every ingredient in a favorite recipe. At a somewhat more complex level, proportions might change in an architectural model so that the ratio of a room's length to width remains 5:3. No matter the size of the room, the relationship of the dimensions remains the same.

The mathematics of proportional reasoning involves the key idea of generalization. Making mathematical generalizations is something that middle-school students often find difficult. It involves thinking beyond individual numbers and how numbers are operated upon, to thinking about numerical relationships and using these relationships to make predictions. Without the ability to generalize, high-school and college mathematics are not accessible. Given the critical nature of this skill, what mathematical challenges can science centers create that will engage visitors in proportional reasoning and in making generalizations?

Almost any exhibit or program that involves the functioning of simple machines can be used to involve people in thinking about proportions. The physics of work is based on machines like pulleys and gears that enable one to systematically vary the amount of effort put into moving something over a given distance. Gears are one form of simple machine found on the floors of almost all science museums.

Unfortunately, most gear exhibits do not explicitly capitalize on the amazing mathematical relationships that they embody. What could be done to mathematize the typical gear exhibit and make it incorporate more proportional reasoning? As part of the Math Momentum project, we created a set of gears that could be used to engage visitors in proportional reasoning. Visitors are challenged to use gears to come up with stories about when two elevators will meet. A photo of the gear set is presented here.

The first step in working with the gears is to encourage visitors to explore gear wheels of different sizes (e.g., those with 10, 20, 30, 40, and 60 teeth). This experimentation takes place within the context of using the gears to move two different elevators up and down. Using the set-up, visitors can address questions such as these: How far down does the elevator go when the biggest gear is turned one complete revolution? How far is the distance the elevator travels when the smallest gear is turned? Because there are several gears with different numbers of teeth, people can explore the relationship between size of gear and distance the elevator travels.

After exploring the gears, visitors are presented with this mathematical challenge: Place elevators at different starting

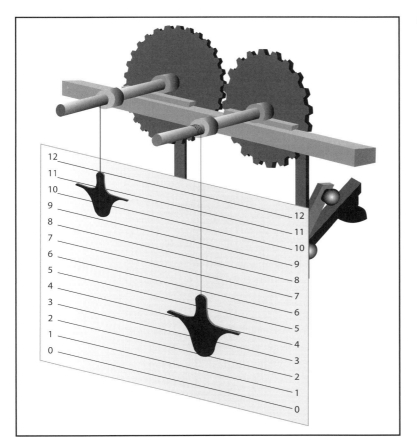

points and determine when they will be on the same floor. Hide the gears so that people observing the movement can see only the floors where the elevators started and the floors where they meet. Their task is to determine the size of each gear (or the relationship of the sizes).

However, rather than presenting the challenge so starkly, it is given in the context of making up a story about the meeting of the two elevators. Storytelling is an excellent way into math for many individuals because it helps them understand what the mathematical task involves. Here is an example of one story that was constructed in conjunction with the gears challenge:

Octavia and John-John are both star soccer players, and they are staying in different ends of different levels of a large dorm with an elevator at each end. They want to pass the ball to each other, and need to be on the same floor to do so. They start out two floors apart, with Octavia on floor 12 and John-John on floor 10. When both gears are turned the same amount, John-John is now on the 7th floor and Octavia, on the 8th. With the next turn, they each end up on the 4th floor, and John-John makes the pass to Octavia, who receives the ball. The question is which two gears could have been used to make the elevators work this way?

The data are as follows:

Octavia	John-John
12	10
8	7
4	4

When Diane Miller and Robert Powell (Saint Louis Sciencenter) presented this story at a workshop for science-center staff, people began thinking about the ratios. Loren Stolow (Museum of Science, Boston) responded to the challenge by noting, "For every four floors that Octavia changed, John-John changed by three. The ratio is 4/3. So there had to be a gear of 40 and a gear of 30." Others noted that what was important in the story was not the number of the floor where the two characters met (e.g., the 4th floor), but the *changes between floors* and comparing these changes for Octavia and John-John. The important mathematics is that described by Loren—every time Octavia goes down four floors, John-John goes down three. This generalization allows you to predict where Octavia and John-John would be on the next downward turn. In fact, it allows you to predict Octavia's position by knowing John-John's (or the reverse) with respect to any floor. Quite a powerful generalization, and one that is essential to algebraic thinking!

The gears activity has been used successfully in several science centers, in each case as a mediated activity involving a staff person. Wherever it has been tried, we have found that the motivating combination of storytelling and movement within this mathematical challenge is a strong way of engaging people with math. Visitors have opportunities to develop characters and plot, and to use drawings and stickers to illustrate their story and their mathematical points. TERC researchers (Noble *et al.*, 2001; Nemirovksy and Wright, 2004) have found that children are able to understand the mathematics of change much more effectively when they link mathematical changes to changes in a storyline.

Time machine

Dealing with very large or very small scales is always a challenge, and can be made into a meaningful piece of mathematical work. The science of astronomy uses huge distance scales, and often makes use of powers of 10 and the notion of light-years to help us get a handle on these distances. We are familiar with science-center models of the solar system that are to scale, and almost always the distant planets need to be off-site (in some cases miles away) to accurately represent the scale. (Fortunately, Pluto is no longer viewed as a major planet. At the other extreme, the emerging field of nanotechnology deals with engineering very small objects. A nanometer, which is one billionth of a meter and many times smaller than the width of a human hair, is difficult to conceive of, even by adults with a background in mathematics.

How can visitors be helped to understand, let alone measure, objects and distances that are either extremely vast or so small that they can't be seen? In mathematics education, children are often introduced to the idea of "benchmarking," that is, establishing through practice a clear idea of how long a foot is, what a pound feels like when lifted, how much liquid is in a pint, and what a square yard looks like. But none of these benchmarks is very useful when it comes to very large or very small scales. The problem is that in science (as opposed to everyday life) making sense of these large and small time scales is crucial. What can be done to help visitors make sense of how large and small scales are used in science?

At the Buffalo Museum of Science, a key experimental gallery is themed on the Hiscock Ice Age dig site in western New York. The field of paleontology is very exciting to both children and adults, but it is hard to conceptualize the huge time scale that is involved in the evolution of the earth and its flora and fauna. In many science centers, including the one in Buffalo, the paleontology exhibits feature a time line showing the relative lengths of different ages. But these time lines do not make sense to many visitors, who are used to thinking of time in terms of their own lifetimes, at the most. For example, staff at Buffalo frequently heard children wonder about whether their parents were alive during the time of the mastodons. And the mastodon era and the era of the dinosaurs seem indistinguishable to many visitors. After all, both were really, really long ago! How does one help visitors distinguish among hundreds, thousands, millions and billions of years?

A kinesthetic time machine.

Staff at the Buffalo Museum of Science experimented with helping visitors use movement to understand time. This was done through a simple time machine consisting of a cranking device and a series of spirals to represent the amount of time traveled.

Visitors were asked to crank back to the time when they were born (assuming one crank per year, a few cranks would do the trick for most children), then to crank back to the time when their parents were born. At this point, it makes sense to switch time scales, for example, by executing one crank per generation. This would enable, with a reasonable amount of effort, cranking back to the time to about 10,000 years ago, roughly the time of mastodon extinction. Ten thousand years represents about 400 generations, assuming 25 years to a generation. Although it takes a fair amount of cranking to go back that far, it can be done in a few minutes—a small expenditure of energy compared to that needed to go back to the time of the dinosaurs. According to many paleontologists, dinosaurs became extinct about 65,000,000 years ago. We'll leave it to you to figure out how much cranking it would take to go back that far! (*Hint:* It will take 6500 times longer than it takes to go back to the era of the mastodon extinction.) Few visitors would have the patience to do this much cranking, and a center wouldn't be open long enough at a stretch for visitors to do this much time travel. Yet those who embark on any of this time travel cannot help but come away with a better sense of when these significant events happened and how much time elapsed between them.

Why is time measurement important from a mathematical

point of view? We've seen that it is important in understanding the sequence and spacing of events involved in evolution and paleontology. It is also important in understanding events that happen quickly, such as some cases of cell division. Mathematically, understanding time necessitates understanding the structure of number systems—including number systems that aren't based on 10. For example, understanding the structure of seconds, minutes, and hours means working within a system that is based on 60. The base-10 system, of course, is the foundation for understanding the relationship between years, decades, centuries, millennia, and so forth.

The time machine, like the other mathematical challenges described in this section, clearly meets the criterion of involving significant math, in this case within the context of paleontology. There are many questions that can be explored, and though there may be definable answers to many of these questions, visitors can choose and explore their own questions. The key to success in this activity is the pairing of mathematical and kinesthetic ideas: Cranking the time machine gives visitors a direct—and sometimes tiring!—sense of relative time scales.

Designing "layered" mathematical challenges: reaction time

In many science centers, a popular hands-on activity involves pressing (or releasing) a button as quickly as possible in response to a sound or light. Visitors are often interested in examining and improving their reaction times. Predictably, they are quite interested in competing with each other to see who is fastest. Unfortunately, in most of the reaction-time activities at science centers, there are no mechanisms to record or compare data. Mathematical opportunities are being missed.

There are several potential mathematical challenges that could be incorporated into reaction-time activities. At the Sciencecenter of Ithaca, an enhanced reaction-time exhibit is being developed to incorporate different levels of mathematical challenges to match the needs and interests of a variety of visitors. In the basic activity, visitors see bar graphs of up to nine successive reaction times. The x-axis shows the nine trials, and the y-axis represents time. The higher the bar, the slower the reaction time. The first mathematical task is to examine the bars and determine which one is best or fastest. This is a basic level of challenge, but it is difficult for many visitors. Seeing that 0.4 seconds is *slower* than 0.3 seconds helps visitors understand that the highest num-

How Quick are You?
Sciencecenter of Ithaca, NY.

ber or tallest graph doesn't always correspond to the best outcome.

The second level of challenge involves setting a target reaction time (or target range), then trying to produce a time that gets as close as possible to the target. This challenge integrates a physical task with a mathematical one: Achieving a reaction time close to 0.5 seconds, which is not an unusually fast time, takes practice. Visitors are challenged to use the feedback from their graphs to get successively nearer to the target. In the process, they learn to read graphs more closely than they would otherwise do. They are learning about mathematical representations of data as they respond to this challenge.

The third level of mathematical challenge in the Sciencenter's reaction-time activity involves learning about the characteristics of a data set, particularly about where the average or mean is located in a data set and what this represents. In this challenge, visitors are asked to come as close as they can to a target (e.g., 0.5 seconds) over the course of six cumulative trials. If one of a visitor's trials were too fast, he or she would need to compensate by going more slowly on one or more subsequent trials. Again, the physical challenge is closely integrated with the mathematical one. But at this level, visitors must also construct a working definition of average, one that involves paying close attention to deviation, or how far away from the average one is on each trial. What matters is how one can balance the deviations over and under the target in order to arrive at an average. This kind of thinking involves visitors in constructing the idea of average for themselves, an important piece of mathematical understanding (Mokros and Russell, 1995).

Constructing the reaction-time exhibit to thoughtfully include several layers of mathematical challenge was a demanding process, particularly because volunteers initiated much of the work. Kathy Krafft, Director of Exhibits, describes the following development process.

Voice from the Field

Developing the Exhibit How Quick Are You?

Kathy Krafft, Director of Exhibits, Sciencenter of Ithaca

Sciencenter staff and volunteers developed an interactive exhibit where up to three visitors test their reaction times by quickly letting go of their push buttons when a light comes on after some random wait time of less than five seconds. A computer-generated display shows bar graphs with individual trials, histograms of up to nine trials, and averages.

Our goal was to create an engaging exhibit that naturally encourages exploration of data and graphing by using meaningful, visitor-generated data.

Exhibit Development

We thought it would be wonderful to have a graph incorporated into our mechanical reaction-time exhibit. From past experience in developing exhibits for a traveling exhibition on math, we found that visitors will not plot out data on a board or chart. We concluded that we would need to develop some automated data collection and display system, probably requiring a computer.

We had also learned that visitors are much more interested in collecting data about themselves, such as their heights.

Finally, we thought that there would be a competitive aspect to this exhibit (trying to be fastest) that would encourage visitors to stay and be engaged in the data collection and graphs.

Several of our key, long-term volunteers got excited by this idea and developed software, hardware, and a cabinet for a permanent exhibit on our museum floor.

An initial prototype was created with two push buttons wired in a shoebox. We found kids really enjoyed the activity and stayed for many rounds, watching the bar graphs. This encouraged the team to advance with the exhibit idea and led to some significant refinements and additions after three more rounds of testing prototypes.

Results

We see friendly competition among visitors; most stay for many trials and explore the trends in their data. Younger kids, adults, and teens—all seem engaged. Visitors requested we include average times over multiple rounds as well. We have also found that visitors enjoy making up additional challenges, such as trying to get closest to a user-selected time or average.

Final Challenges

Throughout this chapter, and the entire book, we present work in which mathematics becomes more relevant, engaging, and motivating. In this chapter, it is clear that creating challenging mathematical activities involves much more than building traditional math problems into existing activities. Instead, math challenges require a foundation that is composed of many ingredients. In all of the examples presented above, math challenges involve truly engaging questions, substantial mathematics, a clear connection with science, problems where the answers are not known in advance, a strong collaborative element, and a physical challenge of some kind. Including as many of these six ingredients as possible will optimize the impact of a mathematical challenge.

The mathematical challenges discussed in this chapter, for the most part, necessitate some mediation. They are either part of programmatic activities or classes, or they involve an explainer working with visitors at an exhibit. Although many of the activities, such as the reaction-time challenges, are effective without mediation, a good explainer adds a great deal to these challenges. A staff member or volunteer is particularly vital in getting the ball rolling by raising questions such as, "What would happen if. . .?" She or he is also vital in promoting collaboration or providing a collaborator for the individual visitor. Because mathematics is inherently a collaborative enterprise—one that benefits from different perspectives—mediation is bound to enhance the mathematical learning that takes place. Because it is expensive to provide this mediation on a consistent level, it is important to figure out ways of using volunteer and staff time optimally to draw visitors into mathematical challenges. Once they are involved, not as much mediation may be necessary.

If presented well, mathematical challenges involve people in exploration that pulls them in and makes them want to keep going. But the success of these challenges has a downside: Most people have not thought about math as a discipline that involves compelling challenges. Some staff and visitors, after becoming deeply absorbed in a challenge such as catapulting or reaction time, don't believe that they were actually doing mathematics. The reasoning is as follows: "Mathematics is boring and frustrating. I just did a challenge that was fun and not too hard for me to figure out, so it couldn't have been mathematics." The important public relations challenge for staff is to use mathematical activities such as those presented in this chapter to convince visitors that mathematics is (among other things) a tool that gives science more depth. It is also a tool that empowers visitors to address the questions that they have about science—the questions that matter most to them.

References

Clements, D. and G. Bright. 2003. Eds. *Learning and Teaching Measurement: 2003 Yearbook*. Reston, VA: National Council of Teachers of Mathematics, Inc.

Lehrer, R. and L. Schauble. Eds. 2002. *Investigating Real Data in the Classroom: Expanding Children's Understanding of Math and Science.* New York, NY: Teachers College Press.

Mokros, J. R. and S. J. Russell. 1995. Children's concepts of average and representativeness. *Journal for Research in Mathematics Education* 26:20–39.

National Council of Teachers of Mathematics, Inc. 2000. *Principles and Standards for School Mathematics.* Reston, VA: NCTM.

Nemirovsky, R. and T. Wright. 2004. Storyline. *Hands On!* 27:2–4. (TERC Newsletter)

Noble, T., R. Nemirovsky, T. Wright, and C. Tierney. 2001. Experiencing change: The mathematics of change in multiple environments." *Journal of Research in Mathematics Education* 32:85–108.

chapter 5

Building Math into Exhibits

Jan Mokros and Ricardo Nemirovsky[1]

Building math into exhibits involves two basic starting points and from there, a myriad of design possibilities. One approach is to mathematize an existing exhibit by finding and capitalizing upon key mathematical ideas related to the exhibit's themes. The main advantage of this approach is that there is an existing foundation for the math work. In addition, it is usually less costly to modify an existing exhibit than to build one from scratch. The second approach is to build mathematics into an exhibit from the start. This method provides a fresh canvas for exhibit designers, allowing them to design their work around carefully chosen mathematical themes. In this chapter, we look at both approaches in the context of mathematical exhibits, as well as exhibits that integrate science and math.

Mathematizing Exhibits

A visitor with mathematical background can recognize the mathematics related to or involved in most science, engineering, and technology exhibits. On the other hand, for someone without such familiarity, the same exhibit might be completely mathematics-free. Mathematizing an exhibit means transforming it so that the mathematical content that an expert is likely to notice becomes more visible to others. Mathematizing may also mean including more explicit math content and processes in the exhibit. For example, you might build measurement challenges (along with measurement tools) into an exhibit on waterwheels, integrate ways of recording and comparing data into a gravity well exhibit, or bring more quantitative elements into an exhibit on

[1] Keith Ostfeld, Andee Rubin, and Tracey Wright provided substantial input to this chapter.

the costs and benefits of "green" construction.

What does it take to mathematize an existing science or technology exhibit? How can math be integrated into an exhibit so that more visitors will recognize and engage in the math? In what ways might mathematizing an exhibit enhance the scientific content of that exhibit? The six ingredients of mathematical challenges described in chapter 4 serve as a benchmarks for the design process and are important to keep in mind when undertaking another fundamental step in planning, a mathematical floorwalk. A floorwalk can help you identify mathematical opportunities, as well as next steps for capitalizing on them.

Mathematical Floorwalks

Exhibit design, whether it involves purely math exhibits or mathematically enhanced science exhibits, benefits from starting with a floorwalk. Selectivity is the key to mathematical floorwalks. It's important to have a focus and to try to achieve depth. Resist the temptation to go through the entire center making a list of all possible mathematical connections. Instead, select a handful of exhibits or exhibit components that offer some intriguing mathematical possibilities. One way to focus is to choose a math concept, such as measurement, and examine exhibits that have the potential to develop visitors' understanding of that concept. Another method is to focus on a type of exhibit, such as exhibits involving simple machines.

Conducting a focused floorwalk involves three stages—planning, implementing, and analyzing—and it involves staff collaboration at all three stages. Throughout this process, it is beneficial to work with the same small team of exhibit designers and education staff.

Planning

The first step is to identify the potentially math-rich exhibits you will examine, with a focus on either a family of theme-linked exhibits (e.g., simple machines) or on a mathematical concept (e.g., measurement). The exhibits selected for the floorwalks need to have the *potential* for mathematical development. If the starting point is a mathematical concept such as measurement, begin by listing all of the places where measurement could play a significant role in visitors' experiences, for example, exhibit components on topics as disparate as dinosaurs, the human body, nanotechnology, and the solar system. If the starting point is theme-linked exhibit components, begin by compiling a list of these exhibits and brief descriptions of each.

Once the list is compiled, the floorwalk team meets to discuss the mathematical understandings that might be linked to these exhibits. Be as specific as possible about the mathematical ideas. Though it is tempting to list one-word topics such as "graphing" or "patterns," these don't mean the same thing to everyone on the team and are too general to be of much use. Rather, try to generate complete ideas and articulate to each other what mathematics could be involved.

In the following example, three related exhibitions were selected for study at the floorwalk: ramps and balls, roller-coaster, and stream table. Of course, there are a few different components to each, as well as several activities that visitors could undertake.

Big Mathematical Ideas for Exhibits Involving Motion

- Measuring "how fast" or "how slow," using clocks and other timers

- Seeing and understanding data that involve the comparison of two or more velocities (e.g., boats at the stream table or cars on the roller-coaster); being able to tell which is faster

- Understanding how to represent velocity, especially by making and using the real-time graphs that accompany the exhibit

- Understanding what it means to "go faster and faster" or "go slower and slower"

- Linking kinesthetic understanding of going faster and faster (or slower and slower) to visual representations of these ideas (possibly at roller-coaster)

This selection of three exhibits is about the right number for a single floorwalk.

It is helpful when generating these ideas to refer to the NCTM Standards (2000), which provide useful outlines of both process and content understanding for students of different ages. The idea isn't simply to match standards to exhibits, but rather to identify key mathematical ideas that could be articulated further as part of an exhibit.

Implementing the floorwalk

Once a team generates a beginning set of math ideas, the next step is to develop questions to guide the floorwalk. These questions should focus on the exhibit itself, the math that is potentially embedded, and also the ways in which visitors might learn about math from the exhibit. For example, below is a set of questions that guided a floorwalk undertaken at the Saint Louis Sciencenter (March 18–20, 2004). This floorwalk focused on the topic of measurement. Small groups each chose one exhibit to examine and took these questions with them to the floor to guide their work:

> 1. Examine the exhibit carefully. What attributes of real phenomena are possible to measure at the exhibit?
>
> 2. How would measuring help you better understand the phenomena in the exhibit?
>
> 3. What do you think visitors could learn about measurement from interacting with this exhibit?
>
> 4. What do you imagine people would bring to this exhibit (experiences, intuitions, etc.) from their lives and backgrounds? What do you bring to it from your own background with measurement?
>
> 5. What ideas does this give you for ways to mathematize the exhibit or create accompanying programs and activities that would increase visitors' understanding of measurement?

As you undertake the floorwalk, take time to observe and analyze each activity. Work with the exhibit yourself, and reflect on your involvement with it. If possible, observe visitors interacting with it, and record what they do and say. Most time should be spent carefully observing the kinds of mathematical learning that might occur and identifying the mathematical potential of the exhibit. One caveat: Because most science-center professionals have a strong creative streak, it is often tempting for them to plunge into modifying an exhibit prematurely. Resist this temptation, and instead focus on understanding the potential of the exhibit.

Analyzing the floorwalk

Analysis involves concentrating on each exhibit, in turn, to discuss the mathematical ideas that were uncovered. Discuss what was seen and how it relates to a key mathematical idea. The mathematical ideas revealed through the floorwalk may be different from the ones that were initially brainstormed. For example, here is the beginning piece of analysis by a group that examined a seismograph:

> In this exhibit, you jump on a platform, creating rhythms and peaks on a real-time seismograph read-out. We liked being able to see the graph and how it changed in response to our motions. Connecting what we did with the visual of the graph was a good way to understand what a graph represents over time. But the *y*-axis of the graph was labeled with "microns." Nobody in our group knew how big they were. The visitors we observed didn't seem to talk about microns, but a couple of them mentioned the Richter scale. So, we had a lot of questions. How big is a real earthquake? Is it measured in microns? How do microns relate to the Richter scale? How much bigger is an earthquake than the motion we created by jumping on the platform? We were sure it's a lot more, but it would be nice to be able to see whether it's ten or a hundred or a thousand times more. Looking further at measurement units in this exhibit would tell us more and make it easier to understand "How big is an earthquake?" We wondered whether it was possible to introduce the logarithmic scale in the context of this exhibit or whether this would be too much to attempt (group report from analysis of Earthquake Exhibit, Saint Louis Sciencenter, March 18–20, 2004).

Although the group originally identified measurement as a broad mathematical focus area, their floorwalk led them to a more specific idea of how the exhibit might promote understanding of measurement units and the relationships among them. They discussed how the measurement of immediate, easy-to-understand phenomena might be related to the measurement of much bigger phenomena. One consequence of this discussion was the group's examination of other exhibits, including those with very large (astronomical) and very small (nanometric) units. A big question that emerged was "How do we help visitors understand the range of measurement units and how these units relate to each other?" One staff person had a great analogy for this mathematical task. She explained it was akin to understanding how a set of Russian dolls fits together—a set that includes dolls so small that you can't see them and so big that you'd need to travel for years to get from the top to bottom of a doll.

Modifying Exhibits to Make Them More Mathematical

The floorwalk often provides the entry point for modifying the exhibit and also serves to raise a great number of questions and suggestions for next steps. Now the design and implementation work begins in earnest. At this point, it is best to work with a smaller team. A single educator, to ensure the mathematical messages are met, and a single exhibit designer, to help determine best ways to modify the exhibit, should be sufficient staff to work on modifications. Too many staff can slow the enhancement.

The next piece of work involves brainstorming possible modifications to the exhibit, keeping in mind the mathematical messages that could be conveyed. Don't be surprised if the first stab at brainstorming results in a false start. Not all exhibits lend themselves equally well to mathematizing, as the following experience from the Sciencenter of Ithaca suggests.

Once an exhibit has been selected, the next step is to consider the possible ways it could be mathematically enhanced. There are four basic categories of enhancements to consider: graphical, structural, tool- or manipulative-based, and programmatic.

Graphical enhancements come in the form of signs and activity sheets. These are fairly popular, as they are inexpensive and don't require staff time beyond the production of the graphical elements. However, they tend to have the least impact, as not all visitors read the signs. Signs and worksheets tend to lack the educational impact of an interactive experience, but can be an important complement to these experiences.

Structural enhancements fully incorporate mathematics into the exhibit experience. For example, a wingspan exhibit might be modified so that visitors can easily record and display their wingspans (and compare them with those of others). When done well, these enhancements make visitors fully aware of the math and help them tie it into their experiences. However, structural enhancements often are more costly than other modifications. In addition to the costs of renovating or rebuilding the component, there may be a considerable amount of exhibit downtime.

Voice from the Field

Picking an Exhibit to Mathematize

Charlie Trautman, President, Sciencecenter of Ithaca, NY

Initially, our museum's exhibits committee zeroed in on our gravity well exhibit to bring out more math for our visitors. This exhibit is a standard one, having many connections with math. Many museums own the exhibit, so we felt that if we succeeded, we could disseminate our math add-on concept to a wide audience. After much enthusiastic brainstorming, however, we decided to select another exhibit because of the practical constraints of the gravity well. Our plan had been to mount a camera above the exhibit and track on a computer screen the path of a coin rolling in ever-faster circles toward a central hole. We envisioned developing a story around data collection, graphs, challenges between visitors, and other activities. In the end, based on what we observed, we decided there were too many problems. Visitors tend to launch multiple coins; they often reach in and grab coins while they are still rolling; there is no good place to put signage that visitors can easily read; and the cost of writing software to test the idea is high.

Far simpler and cheaper than rebuilding or enhancing the structure of an exhibit is providing additional manipulatives or tools to enhance the experience. Simply adding measuring tapes, counting pieces, pattern blocks, or even chart paper can help visitors experience more mathematics. For example, in exhibits involving race tracks, visitors frequently discuss how they can "beat the record." If a measuring tape is available, visitors can easily compare distances and begin using basic arithmetic operations to determine how much farther one car traveled than another. However, a difficulty with adding these tools into exhibits is that they tend to disappear and need to be replaced.

Programmatic enhancements through facilitated activities provide one of the best methods for imparting the mathematics in an exhibit. By engaging visitors in a facilitated experience that reinforces the exhibit, a museum is capable of having profound mathematical impact. This type of mathematizing involves a greater commitment in terms of staff and volunteer training and in the time they spend implementing the activities. Those who are facilitating the activities need to understand the mathematical messages they are conveying.

Of course, more than one type of enhancement can be selected for a given exhibit. No matter what types of enhancements are chosen, the most important step to take before implementation is prototyping. At this stage, it is essential to see what visitors understand about math as they are using the exhibit. Often, it is tempting to focus only on durability during prototyping; however, if visitors do not understand the message or fail to engage in the expected mathematical behavior, some revision will be necessary. Even after prototyping is complete, ongoing evaluation must be conducted to ensure that the modification truly has mathematized the exhibit. In both prototyping and evaluation of the final product, visitors should be asked what they felt the activity was about and what math they think they engaged in. This may help clarify whether the mathematics message is being presented in a way that is obvious to the public. In addition, to gather data about the extent to which visitors are doing math, it is critical to observe them engaging with the exhibit. Sometimes visitors are doing math without realizing they are doing it. They often believe that math must relate to numbers and don't recognize that some skills, such as finding patterns, using logic, and making visual or kinesthetic comparisons, actually involve important mathematics. A combination of interviewing and observing is the best way to determine the extent to which the exhibit has been effectively mathematized.

Voice from the Field

Altering Exhibits to Make Them Mathematical

Keith Ostfeld, Director of Visitor Experience, Children's Museum of Houston

The Children's Museum of Houston began its mathematizing process quite innocently. To ensure that we were providing the best possible learning environment for a very diverse group of learners, we developed an evaluation tool based on Howard Gardner's theory of multiple intelligences (Gardner, 1999). In 1998, we conducted our initial floorwalk, examining each of our exhibits to determine our strengths and weaknesses across all the multiple intelligences. One of our weakest points, we discovered, was in logical-mathematical intelligence.

At first very concerned, we conducted a more careful analysis of our galleries. We determined that there was mathematical information in our exhibits, but we weren't bringing it to life. How could we bring out the mathematics, given that we didn't have the time or money to completely recreate all the exhibits in the museum? We decided to try some enhancements.

First, we decided to make what changes we could to the exhibits. We wanted to maintain our high level of visitor experience, yet provide changes that we hoped would allow our visitors to "see" the math. Most commonly, this was through signage and other graphics work. For example, we developed a sign in our shopping area that we hoped would prompt visitors to discuss sales tax. In our Bubble Lab, we added a take-home sheet that encouraged visitors to go home and mix up their own bubble juice. It then prompted them to try changing quantities of the ingredients to see what would happen. Throughout the whole museum, these little enhancements began to appear.

Although some families did utilize the enhancements, the majority of our visitors either didn't realize they had occurred or didn't find them attractive enough for engagement. Parents may have read the signs, but they didn't know how to put them into context for their children. Occasionally, visitors would engage in mathematical behaviors, but not to the extent that we desired. We wanted to know how to get the same mathematical fervor that math exhibits had engendered into the rest of our galleries.

Later in our work, after we had been immersed in thinking about how visitors learn math, we decided to give exhibit enhancement another try. We conducted a new set of floorwalks, using what we had learned. Then we tried different enhancements. In some cases, we improved the graphics. In many others, however, we provided additional manipulatives designed to help bring out the mathematics. We added rulers, counters, or any other manipulative we felt the public could use to help address the mathematics in the exhibit. For example, we added a measuring tape to a racing exhibit so visitors could keep more accurate accounts of their records as they raced each other and tried to improve their distances.

We also conducted a floorwalk on a single gallery, our Bubble Lab, where visitors are encouraged to engage in open-ended exploration of the properties of bubbles. We still had not been one hundred percent successful in opening up the myriad of mathematical avenues available to our visitors in this gallery. Now, with this new process, our assessment led us to redevelop all the graphics in the gallery, rather than just a single sign, to focus on the mathematics of the geometry and properties of bubbles. We also rethought how to use the gallery and incorporated a set of facilitated activities that further enhance the math, including examining ratios by mixing bubble solutions, gathering data by counting bubbles, and making measurements of diameter and circumference, among others.

Building Mathematically Oriented Exhibits from Scratch

Though building exhibits from scratch can be expensive, many science-center staff believe it is the best way to work, because math is a consistent guiding force throughout the design work. The cost may be higher, but there generally are fewer constraints.

Although a floorwalk is once again a good starting point for brainstorming exhibit ideas, there are other guiding principles to keep in mind when planning a new exhibit that incorporates math. First, remember that mathematical processes—especially communicating, collaborating, representing, and problem-solving—are as important as any mathematical content that will be conveyed. In fact, successful math exhibits involve visitors directly in mathematical processes, or the doing of mathematics, rather than just the viewing of it. Second, keep on the front burner the ingredients and examples of mathematical challenges discussed in chapter 4.

Another starting place for designing mathematical exhibits is provided by the Exploratorium's exhibit model of APE, or "active prolonged engagement" (Humphrey and Gutwill, 2005). Exploratorium developers, like all exhibit developers, have found that building exhibits that deeply engage visitors is a far more complex task than building innovative exhibits (which is difficult in and of itself!). Many of the Exploratorium's APE exhibits are mathematical in nature, and they provide a good starting point for designers of math exhibits. The characteristics of exhibits that foster active prolonged engagement are similar to the ingredients of mathematical challenges discussed in chapter 4 and include the following:

1. Posing challenges that can be solved by doing the activity. Good mathematical exhibits start with a call to action, often a simple challenge such as, "Which of these (carts, wheels, etc.) will get to the bottom faster?" Direct questions like these are effective in prompting both immediate as well as prolonged engagement.

2. Trying to delight visitors, rather than confusing them. Interestingly, visitors aren't always attracted to confounding events, nor do they use confounding phenomenon as a starting point for their own investigations. This may be because they are not very surprised by the things that would surprise a mathematician or a physicist. However, visitors often will observe

simple mathematical phenomena with fascination. For example, they may pay careful and prolonged attention to a Foucault's pendulum that regularly knocks over a succession of dominoes.

3. Encouraging visitors to ask and address their own "what if" questions. Once visitors are engaged and delighted, they frequently ask questions. For example, in the pendulum example above, visitors often spontaneously start searching for patterns and ask questions such as, "How often is a domino knocked over by the pendulum, and is there always the same amount of time between knock-downs?" In this case, delight leads to the profound mathematical activity of pattern-finding.

4. Ensuring that there are multiple intriguing outcomes. One way of engaging visitors is to encourage them to keep exploring. Provide them with activities that result in many intriguing outcomes.

Building a Science Exhibit That Incorporates Math

Because new science exhibits are always being designed, there are many opportunities to build math in from the start. Math can always support the science in an exhibit, just as it always supports the scientific work being portrayed through the exhibit. Nowhere is this clearer than in the case of dinosaur exhibits, which are not only a continuing source of wonder for visitors, but also a platform for demonstrating the innovative ways in which paleontologists use mathematical tools to make scientific inferences. Everything scientists know about dinosaurs they have learned based on measurement and data analysis, although few dinosaur exhibits make this connection.

The Fort Worth Museum of Science and History provides an outstanding example of incorporating mathematical work into its exhibit Lone Star Dinosaurs. Designers began with the goal of having visitors ask the question " How do you know?" for all aspects of the exhibit. Jim Diffily, Vice President and Lone Star Dino curator, didn't want the exhibit to be the common "bone show," with beautifully articulated skeletons, but little sense of the science involved in the displays.

According to Jim and Colleen Blair, Senior Vice-President at Fort Worth, it was easy to include mathematics in the exhibit, because measuring and data collection are integral parts of the science they wanted visitors to engage with. In the Lone Star Dinosaur Exhibit, visitors measure and record data to find out scientific history by looking for patterns in data.

Part of the exhibit design derived from the front-end evalu-

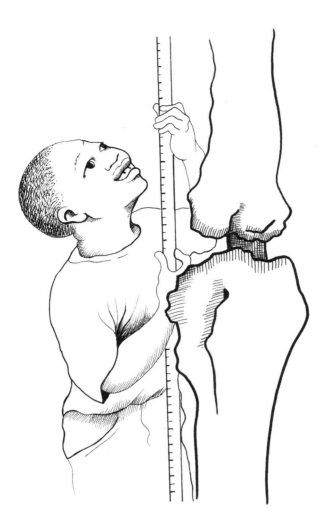

ation study, which found that many Forth Worth Museum visitors needed support in their observation skills when dealing with specimens and artifacts. They didn't have experience in two skills that are critical to both science and math: deciding what to observe carefully and knowing how to find patterns. Therefore, a core interpretive strategy was to help the visitor "learn how to look."

The exhibit is divided into several parts, including a Field Site and a Lab, each of which provides a different kind of mathematical opportunity. The Field Site contains bone beds modeled on those you might find at a dig site. Visitors "join the field team" and collect data on the position of bones and the patterns of rocks found in the area of the bones. They take rubbings of fossils and layers in the surrounding sandstone rock to date the bones. They choose a bone in the excavated bone bed, sketch it, measure it, record its position in three dimensions, and use a compass to determine how it lies with respect to North. Completing these data collection tasks is complicated and requires that visitors use complex instruments. For example, the location of a bone is specified in relation to a 25-centimeter square grid on the floor, using a reference marked by a stake at point (0,0). Visitors record their observations, measurements, and sketches in their Field Notes.

While the Field Site activities focus on measurement, those in the Lab require visitors to figure out what they can infer from data. One set of activities is based on a standard proportional relationship that has been established over a wide range of animals, the ratio between femur circumference and weight that holds for modern mammals. Visitors measure the circumference of a dinosaur femur and type it into a computer, which calculates the weight of the dinosaur. A critical decision for exhibit designers was whether, and for how long, they would display the formula, while the computer was calculating the weight. In the end, they displayed it briefly, but visitors didn't need to read it to successfully complete the activity. Although the formulas weren't interesting to all visitors, the size of the numbers was. Based on the circumference of a dinosaur's femur, the dinosaur weighed about 24,000 pounds—12 tons—close to the weight of two elephants!

A dilemma faced by the designers at Fort Worth, as well as those designing any mathematically enhanced exhibit, is to what extent formal mathematical representations and vocabulary should be part of the exhibit. As pointed out in chapter 4, there's a serious conundrum here: If the math ideas in an exhibit are displayed on signage, some people will be reluctant to go near it, let alone to engage enthusiastically. On the other hand, it's hard to change visitors' images of math if they don't know that the fun experiences they are having are mathematical! Exhibit designers should be aware of both possibilities and the kind of compromises that might be necessary to create an engaging exhibit.

Building a Math Exhibit: The Handling Calculus Exhibit

Sometimes exhibits are designed to focus primarily on math, with science playing a supporting role. One of the most innovative and bold new exhibits where math plays a leading role is the Handling Calculus Exhibit at the Science Museum of Minnesota.

This exhibit makes no attempt to hide the mathematical content, but rather engages visitors with challenges, delights them with kinesthetic activities, and encourages them to ask their own questions. The ways in which this exhibit involves visitors with math have been described in its evaluation and in a research article by evaluator Eric Gyllenhaal (Gyllenhaal, 2006).

Handling Calculus includes eight major exhibits, each of which is open-ended and allows multiple entry points and outcomes. Three of the exhibits focus on motion graphs in which visitors move their own bodies, cars, or sliders on a track and observe the real-time graph of the results. One of these motion graph exhibits, Math Tracks, was designed as an Experiment Bench in which visitors constructed or read stories and moved sliders on tracks to enact these stories.

Stories involved familiar trips, such as a trip to the post office or the trips taken by Little Red Riding Hood (and the wolf!) in the classic fairy tale. Visitors use one slide to make each character move at a certain speed, stop at different times, etc. As these movements are taking place, a real-time graph is being constructed. In this manner, visitors learn to match the mathematical representations corresponding with motions such as "moving at a steady speed" or "going faster and faster." These foundational ideas of calculus are thus made accessible to younger visitors.

In the Math Tracks Exhibit component, visitors can also replay the graph of a "story," using either a preconstructed narrative or one of their own creation. In other words, the graph of position vs. time and the motions corresponding to it can be replayed as many times as desired, each time strengthening visitors' connections between formal mathematical representations (graphs) and physical motion. In a related exhibit component, Road Trip, visitors can choose to make their own simulated trips (complete with rest stops) to cities across the country, and can examine the graphs of these trips. Visitors can also follow the Science Museum of Minnesota Math Crew on a trip they took from St. Paul to Duluth, noting how fast the car was traveling at different times.

Other exhibits in Handling Calculus allow visitors to explore the concept of slope (crucial to understanding the process of differentiation) and the process of integration. Finally, one notable exhibit, Archimedes' Limit, focuses on the idea of limits as applied to calculus. In this exhibit, enclosed in a gazebo, a laser and circular mirror create a sequence of shapes with increasing numbers of sides, moving from triangle to square, pentagon, hexagon, etc., until the shape resembles a circle. Of course, the circle is the limit in this case. Visitors to this exhibit are observing the process involved in Archimedes' Limit in action. Although there is not a physical action called for, there is a great deal that can be learned from observing and reflecting upon this beautiful mathematical process.

In the Motion Graphs Exhibit components, an obvious and critical feature is the posing of kinesthetic challenges to visitors. Enacting a story and linking this story to mathematical representations involves physical, visual, and interpretive challenges. Also critical to these exhibits is the mixture of preformulated challenges (e.g., making the sliders move to correspond with this story or graph) and visitor-initiated challenges (e.g., making your own trip story and predicting what the graph will look like). To what extent did these challenges foster Active Prolonged Engagement?

The evaluation showed that many visitors found the kinesthetic challenges of Handling Calculus to be quite different from the challenges of school-based math.

> By graphing moving bodies and sliders on a track, younger children began to understand motion graphs long before their school curricula dictated. Those still taking school calculus were able to complement their studies, while successful students from long ago revisited an old friend they thought they had abandoned. Finally, some visitors whose relationship with calculus had suffered in the past found balms for their old wounds (Gyllenhaal, 2006, p. 12).

A thesis adopted by the exhibit designers is that kinesthetic experiences involving mathematical representations have the potential to unfold rich connections with formal concepts, which otherwise tend to remain removed and abstract.

Why is kinesthetic involvement, such as moving an object by hand as a means to generate a graph, of special relevance for a mathematical experience? This question, in turn, brings up this other one: How is a mathematical experience distinctive? We can articulate certain qualities that are characteristic of at least some important mathematical experiences. One of them is that in many mathematical experiences, the aspects that matter the most are our own actions and intentions. If, for example, we measure the temperature of water while it is being heated, the fact that the temperature increases up to the boiling point and then stays flat constitutes a property of water under certain conditions, a property that is independent from what we do and expect. We are not doing something that results in the water keeping its boiling temperature constant. On the other hand, if we move the slider toward one end of the track and then keep the slider still, the graph of position vs. time might look like an upward line followed by a horizontal one, similar to the graph of water heated to the boiling point. However, in this case, the flatness of the second section is a direct expression of what we did (i.e., kept the slider still) and what we intended to produce.

For someone who does not fully understand graphs of position vs. time, there may very well be a mismatch between the graph's shape and the intended one. Such mismatches become issues to work on and experiment with. Ultimately, it is the ability to see in the graph itself the actions performed by the person who moved the slide that gives us a sense of fluency and expertise with the graph.

On a broader level, the visitor research on Handling Calculus shows that visitors connected previous memories of calculus with their new experiences. In some cases, the connections they made were to ways of thinking mathematically that they had enjoyed and profited from when they were in school. Clearly, some visitors saw the exhibit as an opportunity to rekindle a delight they had once experienced. On the other hand, Gyllenhaal reports situations like the one of a visitor who "read a label, told his seven-year-old nephew, 'This is calculus, James,' and then walked away" (Gyllenhaal, 2006, p. 7).

A mathematics exhibit, to the extent that it is recognized as such, can mobilize powerful memories influencing what visitors see in the exhibit, as well as making possible the revision of those memories in light of new experiences. Such revisions take different forms. For instance, "when we asked one respondent what she had learned reading the 'Take it to the limit' panel, she said that she learned she had had a really horrible calculus teacher in 1981" (Gyllenhaal, 2006, p. 11). One college student said "he liked these exhibits because he got to understand the concepts, whereas in class he had been so wrapped up in formulas and problem-solving that he did not really understand the ideas behind them" (Gyllenhaal, 2006, p. 10).

The Handling Calculus Exhibit took on difficult mathematical content in a very explicit way and did not ever hide the mathematics from its visitors. Such explicitness makes possible a personal and emotional engagement with the exhibits. It is true that for some visitors the word "calculus" is a dreadful announcement or a threatening warning sign, and the immediate response might be avoidance; such was the case for the college student who refused to work with a computer page that mentioned the word "derivative." However, when exhibits offer entry points allowing for rich explorations with the conceptual resources that people do have, such as body motion, discrimination of shapes, spatial orientation, and so forth, the possibility arises that visitors, instead of being locked up by their memories of math, might revise and question what these memories tell them.

The Handling Calculus Exhibit provides an important existence proof concerning the ways in which exhibits can engage visitors in meaningful, prolonged mathematical activity. There is no more important goal for a mathematically oriented exhibit than helping broaden and enliven popular images of what mathematics is and how closely related mathematics can be to the most pervasive activities of daily life, such as talking, walking, moving, and drawing. Showing math in a new light, helping visitors see themselves as capable of engaging in mathematics, and making it possible for visitors to find delight in mathematics are worthy goals for exhibit developers.

Next Steps

Whether mathematizing an exhibit that is being built from scratch or enhancing one that is already up and running, it is important to focus on how visitors will learn from the exhibit. There needs to be a plan for evaluating the mathematics that visitors are learning and the extent to which mathematical attitudes and engagement are affected by the exhibit. Anderson's (2001) groundbreaking case studies of mathematics in science centers showed that the best exhibits were ones that focused on visitor learning, rather than on what the exhibit was teaching, and looked closely at the nature of visitors' mathematical learning. At the time when the studies were conducted (in 2000), there were few instances in which the mathematical impact of exhibits had been carefully examined. Fortunately, the situation is changing, as is illustrated by the research from the Handling Calculus Exhibit, as well as research currently underway at the Exploratorium, the Children's Museum of Houston, the Sciencenter of Ithaca, and other science centers participating in the Math Momentum project. The final test of whether an exhibit has been successfully mathematized is the extent to which visitors have become engaged in doing mathematics and in thinking about themselves as capable and interested learners of math.

References

Anderson, A. 2001. *Mathematics in Science Centers*, Washington, DC: American Association of Science-Technology Centers.

Gardner, H. 1999. *Intelligence Reframed: Multiple Intelligences for the 21st Century.* New York: Basic Books.

Gyllenhaal, E. 2006. Memories of math: Visitors memories and feelings about school math shape their experiences in an exhibition about calculus. *Curator: The Museum Journal*, in press.

Humphrey, T. and J. Gutwill. 2005. *Fostering Active Prolonged Engagement: The Art of Creating APE Exhibits.* San Francisco, CA: The Exploratorium.

National Council of Teachers of Mathematics, Inc. 2000. *Principles and Standards for School Mathematics.* Reston, VA: NCTM.

chapter 6

Math, Families, and Science Centers: Opportunities and Issues

Andee Rubin

The father picked up the golf ball *and a darker ball of the same size from the Weights & Measurements Drawer and placed them on the two pans of the balance. The golf ball was much heavier, even though the two balls looked very much the same. His four-year-old daughter looked confused and asked, "Why are they different?"*

"Pick up both balls in your hands," her father told her.

"Oh, this one's a foam ball!" the daughter exclaimed.

"Sometimes two things that look the same weigh different amounts," the father pointed out. "Remember how we have to weigh the fruits and vegetables in the grocery store?"

"There's a big scale at the coffee shop," the little girl chimed in.

"That scale is a lot like this balance," her father explained. "With the balance, we looked to see which ball was heavier. The scales at the coffee shop and the grocery store weigh things and give you the weight in numbers."

Interactions like this *can* take place in science centers, especially in exhibits explicitly designed to encourage and support mathematical interactions among family members. In this example, reconstructed from a parent–child interaction with a Weights & Measurements Discovery Drawer at the Oregon Museum of Science and Industry in Portland, the elements of a productive mathematical discussion are all present: parent, child, mathematical tool, mutual engagement, and an intriguing question. As simple as this combination seems, it can be difficult to achieve, especially in the busy everyday lives of families.

Science centers have a special role to play in fostering children's mathematics learning through exhibits and programs that engage families in mathematical activities and conversations. Although parents want to support the development of their children's mathematical skills and confidence, there are many obstacles to doing so. A fundamental stumbling block to parents' involvement in their children's math learning is their own uncertainty about how to help. Parents have many ideas about how to support their children's learning in literacy but are less sure of what to do when it comes to mathematics (Hartog and Brosnan, 2003). Parents who do use mathematics activities at home typically rely on flashcards or other types of drill. Understandably, parents are likely to focus on the content and pedagogy that was at the core of their own school mathematics experience (Civil, 2001; Civil and Andrade, 2002).

Research has shown that school-age children benefit when parents support their math learning outside of school. These children may be more likely to achieve academic success, have positive attitudes about learning, and acquire new skills (Scherer, 1998; Barton et al., 2004). There are many programs that promote parental involvement in children's math learning, and homework and homeschool curriculum connections are typically the focus.

However, many working parents express the desire that the limited time they have with their children be focused more on the social and emotional side of family life than on homework (Kralovec and Buell, 2000; Farkas et al., 1999). Here is where science centers can play a leading role. Science centers are places where a family can come as a group,[1] be playful with math and science, work together or apart, ask or answer questions, talk or be silent. However, many children and adults do not see science centers as environments in which to do mathematics. Part of the role of the science center is to provide experiences that can change visitors' minds by expanding their view of mathematics and supporting the development of their mathematical self-confidence. Because more than half of science-center visitors are in family groups (Korn, 1996), there is plenty of opportunity to influence their relationship with mathematics.

Three Themes Regarding Families, Math, and Science Centers

This chapter describes three issues that affect the ways in which science centers can engage families in mathematics through exhibits and programs.

[1] By family, we mean several intergenerational, related people, including but not limited to parents, grandparents, siblings, aunts and uncles, and close friends.

1. Expanding families' images of mathematics and their confidence in thinking mathematically

2. Extending visitors' mathematical experiences outside the museum

3. Making math accessible to a wide range of family visitors by engaging their strengths and interests

In discussing all three themes, we have chosen to focus on the mathematics of measurement. Several science centers have found that measurement is an accessible activity for all families because it enables them to connect their everyday activities to mathematical reasoning. A significant advantage of measuring as a mathematical activity is that it is a crossover process, critical to tasks from cooking and woodworking to bioscience and nanotechnology. This relationship encourages families to reframe everyday tasks, both inside and out of the home, as mathematical (Schliemann and Carraher, 1992). In addition, measurement is the mathematical area in which there is the greatest gap between white students and students of color on the National Assessment of Educational Progress (NAEP) (Lubienski, 2003).

Expanding families' images of mathematics and their confidence in thinking mathematically

Virginia Thompson, a math educator whose primary interest is providing families with positive mathematics experiences outside of the classroom, states,

> The single most important thing we can do to enact reform is change negative messages about mathematics. Children need to practice, think about, and do mathematics outside the classroom to really learn and understand mathematics concepts, ideas, and applications (Thompson, <www.enc.org/focus/family/document.shtm?input=FOC-000729-index> (June 23, 2005)).

Thompson uses reading as an analogy to drive this point home. If the only exposure a child received to books and reading were in the classroom, how would he or she ever come to comprehend and appreciate literature? Just as reading is an activity that transcends classroom boundries, so too is mathematics. The dining room table and the desk in the bedroom are crucial sites where perceptions are changed and reinforced. Because of this, it is essential that families be empowered to extend mathematics learning into the world beyond the school building.

But if parents regard math as the subject they liked least in school (as is the case for more than 50% of the adult population [Burns, 1998] and a topic that only "other" people understand, little will happen outside of school. Science centers have the potential to create environments and experiences that can overcome such personal barriers and math phobias. One approach science centers have taken is to take advantage of the mathematics involved in "real" science, specifically measuring and understanding data. Below are two family-focused examples of how science centers included opportunities for visitors to integrate math and science by measuring and recording data of their own.

Forth Worth Museum of Science and History's Lone Star Dinosaurs Exhibit: How do you know?:
The Lone Star Dinosaur Exhibit at Forth Worth Museum of Science and History is described in some detail in chapter 5, Building Math into Exhibits, as an example of the impact of mathematizing an exhibit from the beginning of its design. One of the primary goals of the exhibit, in fact, was to help visitors understand the ways in which scientists collect and use data in constructing scientific history. The exhibit supported visitors in actually collecting and analyzing data to answer the question "How do you know?" Visitors used tape measures, compasses, and laser tools to make measurements and record them in their Field Notes.

How did families interact with and respond to the mathematics embedded in the Lone Star Dinosaur Exhibit? The evaluation, carried out by Randi Korn and Associates (2005), found that most adults and all children enjoyed using measuring tape and compass, sometimes measuring more than was required by the Field Notes. They especially enjoyed measuring in an authentic-looking dig site. Often this measuring activity was a family affair, with parents helping children use the tools correctly and record what they found. In fact, the children who used the Field Notes most extensively were usually accompanied by adults who had encouraged them to do so. This exhibit (along with several others described here) provides evidence of measurement's great potential for supporting families in doing mathematical work together.

In spite of the enthusiasm displayed by children and adults, many of the children had trouble using the measuring tools correctly. Originally, the exhibit designers had included written instructions for making measurements, but found that they were not sufficiently helpful. Instead they created several videos which

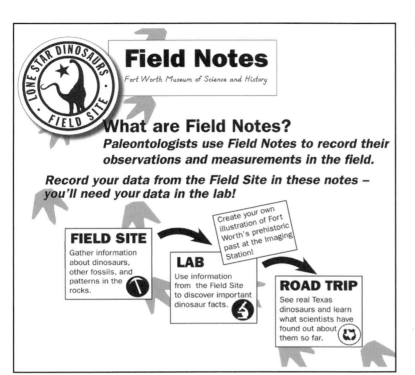

showed children using the tools appropriately. Yet problems remained. Children sometimes didn't know where to place the zero on the measuring tape or how to read the final measurement. The design of the measuring tape confused some children. The exhibit designers checked with the local schools to find out what kinds of tape measures students were using so that they could support children's formal learning as much as possible.

In addition to using the measuring tools correctly, some visitors had trouble making certain measurements because they didn't know exactly what part of the object to measure. For example, some were unsure how to measure the length of a bone. Did the length include only the shaft? Knowing which attribute to measure is at the very core of measurement and an issue that scientists wrestle with constantly. In school, students often know ahead of time what they are going to measure and which measurement tool they are going to use. An advantage of working in an informal context is that it provides students with a situation in which many different measurements are possible, and therefor there is a need to evaluate the appropriateness of each.

The exhibit successfully helped both adults and children to

understand the role of measurement in science. Many reported that measuring is part of a scientist's job. Some said that they were measuring in order to figure out the dinosaur's size; others were less specific and said only that they were learning about dinosaurs. "Getting your hands on the tools" (Randi Korn and Associates, 2005, p. 9) was a reason visitors cited for "feeling like a scientist" when they were at the exhibit; others cited the process of observation and recording. For many visitors, there was a clear connection made between the math of measurement and the scientific process.

Of course, the transition between the measurements visitors make and the conclusions they can draw requires inference, which is often more difficult than the measuring itself. In the Lone Star Dinosaur Exhibit, designers often had to make a judgment regarding how much mathematical inference to make explicit for the visitor. As described in the Exhibits chapter, this dilemma arose in the context of inferring animal mass from femur circumference: How visible should the proportional reasoning equation be? In this case, a compromise appeared to work; the equation was displayed for those who wanted to think with it, but visitors' understanding did not depend on the formal math.

The "naming the math" dilemma occurs in many exhibits. If an exhibit is easily recognizable as math, some people, both adults and children, will be reluctant to go near it, let alone engage enthusiastically. On the other hand, being able to identify a fun experience as mathematical is the only way for visitors to expand their definition of math! The following examples explore this question further.

Oregon Museum of Science and Industry: Discovery Drawers: What kind of mathematics might be appropriate for families with preschool children in a science center? That's the question Ann Wieding of the Oregon Museum of Science and Industry took on in creating early childhood Discovery Drawers. Parents of preschoolers have few ideas about the kinds of math that are appropriate for their children. The image of "math as arithmetic" leaves little of mathematical interest for children to do before they begin school except, perhaps, counting. As in the Lone Star Dinosaur Exhibit example above, Ann faced the task of expanding people's definition of mathematics— and again, measurement was an appealing and appropriate way to achieve that objective.

The Discovery Drawers project was originally designed in the context of science, with the goal of giving parents the tools to become their children's first teachers. The first drawer dealt with butterflies and bugs; a later one, with dinosaurs. Ann and her team then extended their project into mathematics, creating several math drawers and highlighting the math in some of the science drawers. Measurement turned out to be one of the most natural aspects of mathematics to include in many of the drawers.

Ann found several resources particularly useful as she incorporated math into the Discovery Drawer project. Two particularly important sources of information were *Family Math for Young Children* by Grace Davila Coates and Jean Kerr Stenmark and *The Young Child and Mathematics* by Juanita V. Copley (both available through the National Association for the Education of Young Children). A consultant, Swapna Mukhopadhyay at Portland State University, served as a sounding board for Ann's ideas about math for very young children. In the course of these preparations, Ann refreshed her memory of algebra and geometry and increased her awareness of how math and science overlap.

Discovery Drawers are available for families to pull out and use as they wish. The Weights & Measurements Drawer that Ann developed, shown here, has been particularly engaging for families. It includes a scale calibrated in ounces; a pan balance scale; several objects that look similar but have very different masses, i.e., a golf ball, a nerf ball, a pumice rock, and a heavy rock; and a large plastic tape measure, appropriately named "Big Tape Measure." The drawer also contains four books: *Inch By Inch; Math Counts: Weight; So Big! My First Measuring Book;* and *You Can Use A Balance.* The scale is particularly popular, and parents are comfortable with it, as it resembles tools they use themselves.

Ann considered what measurement might mean in the context of the youngest visitors. She created a Numbers & Counting Drawer that contains such things as a bunch of small teddy bears in different sizes and colors; a set of six plastic bowls for sorting; and activity cards that emphasize counting, sorting, and matching. The contents of this box represent Ann's view that measurement encompasses far more than just using formal tools. For young children, counting, sorting, and comparing are forms of measurement and can be natural mathematical experiences in and of themselves.

In addition to the tools in the math drawers, there are several informal ways in which visitors can measure length. Near the door to the Discovery Lab, where the drawers are housed, is a large vertical ruler children can use to measure their own heights and compare them to the heights of other family members. With

Weights and Measurments Discovery Drawer from the Oregon Museum of Science and Industry.

the many tape measures that are freely available, children also measure the tables and chairs, the aquarium, the couch, and just about anything else they can get their hands on.

Each Discovery Drawer contains an activity card with suggestions for tasks that families can work on together, but informal observations indicate that visitors often create their own activities, based on the structure and preferences of their own families. Although the exhibit is primarily designed for children aged 2–6 the drawers are designed to involve other family members as well. Older siblings aged 7–9 may read the activity cards. Even children under two can be actively involved in the sorting activities. Groups of girls have been observed playing school, using objects from several drawers and designating one child as a teacher. Fathers, in particular, seem to be comfortable with the Weights & Measurements Drawer.

The Discovery Drawers, like the Lone Star Dinosaur Exhibit, give families access to and experience with measurement in a way that supports both adults and children. In terms of tool use, however, there are significant differences between these exhibits. The range of tools visitors can use with Discovery Drawers is wider, and there are many options for measuring. So, if measuring length with a tape measure is problematic, children and adults can use the vertical ruler. Families also have the opportunity to work with scales and with counting and sorting, which are simpler tasks.

Discovery Drawers do not attempt to make a connection to formal math, nor are they labeled as mathematical. However, there is no attempt to hide the purpose of a box. Some visitors may make the connection between measuring and science, especially because the themes of many of the drawers are scientific, but that is not a specific goal. Although much of the case for measurement in science centers is based on the role measurement plays in science, it's important to remember that kids enjoy just playing with measurement tools, comparing their heights to others', figuring out how to measure large objects like dinosaurs, and dealing with puzzles like why objects of the same size may have different weights.

We've seen two examples of how measurement activities can expand families' images of mathematics beyond the dry and intimidating subject they remember from school. Because measurement is an integral part of doing science, families engage with measurement in a meaningful context. And, as was seen with the Discovery Drawers, measurement can be presented as a process that can be fun in and of itself, as it supports collaboration among family members.

Extending visitors' mathematical experiences outside the museum

Enabling visitors to think mathematically during a visit to the museum is a significant challenge. How can we provide guidance for families to continue to do math when they return home? In some cases, this can be accomplished by extending families' experiences with an exhibit. In other cases, it is done by setting up stand-alone family mathematical activities within a science-center or community context and encouraging the use of these activities at home.

Fort Worth Museum of Science and History: Extending an exhibit: Designers of the Lone Star Dinosaur Exhibit wanted families to leave with a resource that would extend their museum experience, so they developed a take-home Activity Guide, a part of which is shown here. The focus of the guide is to support families in continuing to do real science and math in their everyday lives. The following two activities are included in the guide:

Will the dinosaur fit in your backyard? Figuring out where dinosaurs of different sizes would fit in your backyard using strings to equal the dinosaur lengths, which vary from 4 to 57 feet.

A very HANDy measuring tool. Experimenting with using body parts or other nonstandard measures (e.g., toasters). "If you could stack enough basketballs, how many would it take to

Lone Star Dinosaurs Family Activity Guide from Fort Worth Museum of Science and History. Laurel Aiello, illustrator; Julie Steele, author.

Lone Star Dinosaurs Family Activity Guide

Fort Worth Museum of Science and History

Your whole family will dig science when you make more discoveries together in your own field site at home!

Lone Star find: One summer, 7-year-old Thad and his father were exploring a dry creek when they spotted a new pattern, unlocking the clues to dinosaur bones! Look more closely at the world around you, and who knows what you'll find!

reach the ceiling?

In the next "Voice from the Field," Julie Steele, writer of the Activity Guide, describes her design decisions and how they were influenced by an increased emphasis on math.

The guide also takes into account the different times when and places where families may have the opportunity to think scientifically or mathematically. One activity is meant to be done on the way home from the museum; some can be used indoors and others, outdoors. By including a variety of possible times, places, and contexts for families to do math and science, the guide extends visitors' interactions with the exhibit.

While the Lone Star Dinosaurs Guide was meant primarily for families to use at home as an extension of an exhibit, the next example involves mathematical activities that are done by parents and children in many contexts, including science centers, community programs, schools, and at home.

Family Math: Family Math originated at the Lawrence Hall of Science because of the concern of an elementary school principal in Richmond, California, who had observed that many parents wanted to help their children with math but didn't know what to do. In response, math educator Virginia Thompson started the Family Math program, which provides math materials designed

Voice from the Field

Interview with Julie Steele, Materials Developer, Fort Worth Museum of Science and History

This interview was conducted in August 2005 by Andee Rubin.

What were your overall design criteria for the guide?

Julie: *The guide is an invitation to families with children aged 5–11 to take the experience of "doing science" home with them and using those same skills in their everyday lives.*

It is meant to be a playful piece that will promote interaction and dialogue about what our observations and data tell us and how we know. Connections to exhibit experiences are meant to reinforce those skills and to provide further examples.

The language and voice of the guide were intended to model the language and tone for dialogue among family members. There are prompts that provide a starting point for a larger discussion about how they, as scientists, can make sense of the clues they observe.

What design decisions were most difficult to make?

Julie: *The biggest challenge was writing a succinct piece that provided compelling, clear examples for families to try at home. The final guide is only one folded sheet, four pages in all. Translating the exhibit activities into familiar scenarios took some effort, and I needed to think carefully about how to include references to the exhibit.*

Were design and revision issues around the math content different from those for science?

Julie: *The math is so well integrated into the sci-*ence content of the exhibit that finding natural places to highlight the math was really a non-issue. As a science educator, it was slightly more challenging for me to recognize the places where the math concepts could be made more apparent with a subtle change in language or syntax. For instance, one activity prompt was originally written as, "How many toasters would you need to line up end-to-end to* make *the length of your table?" In the final version, the question reads, "How many toasters would you need to line up end-to-end to* equal *the length of your table?" It is a slight change, but modeling the use of the vocabulary helps to connect the dots between the math and the science.*

How did research about children's understanding of math influence the development of the guide?

Julie: *It is important in any development process to understand which concepts are familiar and which might be misunderstood. I based some of the content development on research that suggested that children have little concrete experience with measurement. Not only did this encourage me to focus on measurement, but it also made me think about how to construct activities so that family members would have entry points for dialogues about measuring objects, size, or scale. In the end, we hoped kids would have experience using real tools, with knowledgeable adults as guides.*

for parents to use with their children at home. The philosophy and guiding principles of Family Math, as stated on the Family Math Web site (www.lawrencehallofscience.org/equals/aboutfm.html), are as follows:

✓ Family Math believes that all children can learn and enjoy mathematics. Parents and other family members are their children's first and most influential teachers. Yet many parents report that they do not know how to support their children's mathematical learning.

✓ Family Math focuses entirely on families learning mathematics together. In Family Math, mathematics becomes a challenging and engaging learning experience for everyone. Its math topics connect to the school curriculum, including algebra, probability, statistics, estimation, logic, geometry, and measurement.

✓ Family Math focuses on developing problem-solving skills and building a conceptual understanding of mathematics with hands-on materials. These materials are often found in the home.

A typical Family Math activity includes a rationale (the mathematical content), an indication of the number of players and materials needed, and directions for playing the game. Perhaps most important are instructions on how to extend the activity so that it involves problem-solving beyond the original game. Most parents, even if they are comfortable doing math themselves, have little experience talking with their children about mathematical thinking beyond arithmetic problems; this aspect of Family Math helps all families enrich their mathematical conversations.

For example, Family Math includes a variety of strategy games similar to the game of Nim. In one of the many versions of Nim, play starts with an arbitrary number of coins in a row. Players alternate picking up one or two coins on each turn. The person who picks up the last coin(s) wins. Figuring out who will win by making the last move involves considerable strategizing and "if–then" thinking. A Family Math-type activity sheet for Nim would first describe the relationship of Nim to logical reasoning and number theory, then list the rules of the game, and suggest that parent and child play the game several times. It might have a list of the kinds of questions parents could ask while the game was going on. Then, for those who want to go further, the activity description would suggest that parent and child work together to find a general strategy that would enable them to win the Nim game every time. Finally, the activity sheet might suggest related activities that can extend the experience, perhaps involving more family members.

Most of the time, families learn about Family Math through a class that involves parents and children doing math together, so that by the time they leave, they have already had an opportunity to work with one another on a mathematical activity. Family Math is used in many contexts, including schools and a variety of community programs. Family Math gives science centers a vehicle with which to involve parents, especially those from low-income and underrepresented groups, through their already existing family programs and through their community partners.

At Lawrence Hall of Science (LHS), where Family Math was developed, the program is offered in both of these ways. At the center itself, Family Math is offered in the context of early childhood classes attended by parents and children together. At times, the LHS team has had trouble recruiting parents, especially those who are uncomfortable with their own math experiences. However, once these parents come, they tend to come back. Parents find it helpful that the program starts with topics they remember as important parts of math, especially work with numbers. After a while, it's easier for them to move on to parts of math that are less familiar, but equally important.

Lawrence Hall also reaches out to underserved families in community centers through Family Math. Grace Coates worked with "The Kids' Breakfast Club," providing a Saturday morning group for parents who wanted to know how best to help their children with math. Parents learn the activities and are encouraged to use them with their children for at least 30 minutes during the week. The effect on the parents' math confidence can be dramatic, and as they continue to feel more competent, they become better math role models for their children.

In a different approach, Saint Louis Sciencecenter (SLSC) made extensive use of Family Math with community partners, as described in the next "Voice from the Field."

Diane Miller chose to include math in her community program because she thought the Family Math materials did an excellent job addressing the critical issues of inclusion, class, and literacy. The original opportunity to reach parents in the community actually came through the schools. The Sciencenter had been working on a small scale with the local school system to explain

Voice from the Field

Making Math Familiar and Comfortable for Families

Diane Miller, Vice President, Community Science Programs and Partnerships, Saint Louis Sciencenter

Math is scary to many families, often because they don't know what mathematicians do. Frequently, they think that math is calculus and that it has been hidden away where only "smart people" can get to it. In fact, when people know more about math, they don't find it intimidating at all.

Family Math has provided an opportunity at the Saint Louis Sciencenter for a wide variety of families to explore and demystify math together. All members of the families are engaged with the games—and if kids win them, they think of themselves as mathematicians. Now they know some-thing about what mathematicians do, and it's not so unfamiliar after all.

Many of the teens who attend our programs are comfortable with science, but not mathematics, and have ambitions to become scientists who never use math! I've shown them material from Lawrence Hall that indicates that, on average, the more math you've learned, the more money you'll make—but still math seems to be a significant barrier. We hope that the early experiences visitors have with Family Math can prevent them from falling into this trap.

the power of Family Math and to offer their services in leading workshops. As a result, the St. Louis Urban Systemic Initiative schools adopted Family Math, offered by the SLSC, as its required family outreach component.

To implement the school-based initiative, SLSC staff trained parent liaisons (who were already part of the homeschool connections the public schools had established) to give Family Math workshops. From the beginning, Diane and her staff involved the parent liaisons in planning and debriefing the workshops so that both the SLSC staff and the parent liaisons could learn more about mathematics and the processes of teaching and learning math. SLSC staff developed an eight-session Family Math course that culminated in a visit to the science center. When parents arrived at each session, there were several math stations where they could play math games. Rules for the games were written in simple words on large pieces of paper so that parents could figure out how to play on their own. The staff followed these games with two or three guided activities and gave parents collated booklets of instructions for the games they had just played.

SLSC staff always made sure that parents knew that Family Math activities were not just for fun, but also to extend children's math learning and their ability to pursue careers in math and science. Parents seemed to take this to heart. Family Math leaders noticed that parents tended to try the activities at home with other siblings or family members, and felt more confident about assisting their children with homework. Current work on family programs underscores the importance of parents actually playing the games themselves in a learning situation. When families are given a set of math-related games to play at home, they are more likely to play those that they learned in an introductory session. They tend not to take the time to read the directions to learn to play other games—even when the directions are accessible and easy to follow (Kliman et al., 2006).

The hope for Family Math materials is that they will be used many times and will fundamentally change adults' and children's images of themselves as mathematical teachers and learners. Family Math can be a catalyst for more powerful connections between a science center and community groups or individual families. In turn, these can result in more diverse visitors coming to the science center and more science-center-sponsored events in the community.

Making math accessible to a wide range of visitors by engaging their strengths and interests

The criteria for family-friendly exhibits developed by the Philadelphia/Camden Informal Science Education Collaborative (PISEC) (Borun et al., 1998) stress the key element of *relevance*, that is, the importance of providing links to visitors' existing knowledge and experience. Yet it is always the case that families—and different members of individual families—are heterogeneous. How can a math exhibit or program for families address the diverse nature of participants, who may have very different levels or types of knowledge and/or experience? This section will describe the Moneyville Exhibit at the Oregon Museum of Science and Industry, where connecting with families' experiences was a primary goal.

The Moneyville Exhibit began in 2000 with three major goals:

- ✓ To encourage economic literacy
- ✓ To promote mathematics learning grounded in real-life contexts
- ✓ To provide an opportunity for family learning

An overarching principle for the exhibit was to take advantage of families' interest in the subject of money to engage them in economics and mathematics. During the initial phase of exhibit development, economic literacy emerged as the primary goal, with math learning as a secondary goal. With input from advisors, the Oregon Museum of Science and Industry team identified two "big ideas" for the exhibit: (1) Money is about making choices; and (2) Understanding math gives you power in making those choices. The project team wanted to create an exhibit in which family members could feel confident and draw on their everyday expertise. They were aware that some families might hesitate to visit an exhibit on economics and math, but front-end research indicated that the topic of money could work as an entry point for families who might not otherwise be attracted to the exhibit.

The story of Moneyville returns us to questions raised in the discussions in chapter 5 of Fort Worth's Lone Star Dinosaur Exhibit and the Minnesota Museum of Science's Handling Calculus Exhibit. An ongoing dilemma for exhibit and program developers is how much to highlight the mathematics. The Handling Calculus Exhibit illustrates how the inclusion of a sophisticated mathematical word need not be a deterrent to families approaching the exhibit and, for some visitors, can be a draw.

Voice from the Field

The Principles Behind Moneyville

Karyn Bertschi, Lead Exhibit Developer, Oregon Museum of Science and Industry

Knowing that we were trying to engage families affected every aspect of our design of Moneyville. At the beginning, we looked for topics with special relevance for families, for instance, Balancing Your Budget, Better Buy (percent discounts), Lemonade Stand (a simulation), and Make a Million (savings and compound interest). We focused on the decisions that families have to make regarding money so the context for the math would be drawn from daily life and be relevant to them. We also hoped this would reinforce the notion that "everybody can do math."

We used criteria from the PISEC Project (Borun et al., 1998) during the formative evaluation to be sure our exhibits were family-friendly. Wherever possible, we designed components to appeal to multiple users and to have the space for families to cluster around them. We also chose a colorful, playful design to help reduce the potential intimidation factor of an economics and math exhibit.

To encourage families to explore thematic areas together, we chose to place activities for different age groups close to one another. For instance, we placed the Kids' Market, a farmers' market role-play area for younger children, near the Stock Market activity, which was aimed at an older audience. Our evaluation showed that these design elements contributed to the kinds of interactions that occurred within families. We overheard lots of lively discussion between parents and children about making choices with money, especially at Balancing Your Budget and at Get Real, a computer simulation of real-life economic choices.

Our choice of math topics was also influenced by our target audience, families with children aged 5–14. For that age group, we chose math strands that were central to the K–8 curriculum: understanding numbers, measuring and comparing, finding patterns, interpreting data, and thinking and reasoning. We also chose advisors who had experience with the Family Math program, including a local math educator who became part of our core exhibit development team.

We knew that we needed to be explicit about the math embedded in exhibit activities, because parents might not immediately recognize the importance of math in daily life. To underline the math, we included a "Math at Work" label to highlight the math strands in each exhibit activity.

The focus on money generally worked the way we had hoped. The summative evaluation found that Moneyville was very successful in our primary goal of engaging families in economics and math learning. Many of the most popular activities were ones in which parents were com-

fortable as facilitators of learning, because these activities were particularly relevant to family experience. The evaluation underscored a strength of money as an exhibit topic: Because adults have a

related math concepts. Because the money and economics strands were so prominent, it was harder to raise the visibility of math. Visitors may have been "doing math" and "talking math," but they may not

Kids' Market at Moneyville, Oregon Museum of Science and Industry.

certain amount of life experience with money, they seemed more at ease as facilitators of both the math and economics learning in the activities than they might have been with other topics. And, of course, mathematics is inherent in money, as a measurement system.

However, the evaluation also showed that we were less successful in reaching our goal of raising visitors' awareness about math. I think we could have been more explicit about the math in each exhibit and in linking exhibits that involved

have been as aware of the math as they were of the economics content.

What we discovered is that most visitors' definitions of math were pretty narrow—limited primarily to activities involving numbers and operations. So, if an exhibit activity obviously involved numbers and operations, visitors were much more likely to say they had used math in the activity. If an activity involved other math skills and concepts, like data analysis, visitors were much less likely to say they had used math.

The evaluation also documented the success of the project in making exhibit experiences relevant to a broad audience. Visitors were seen getting their own money out of their pockets and purses to compare to that in the exhibits, and they mentioned many connections to real life. A teenage boy said, "This helps me practice for survival out there." A seven-year-old girl at Balancing Your Budget explained, "It's so kids can see how hard it is to choose and see how much money their parents are actually spending just to get them things."

Children sometimes noted the connection to math, but only when asked whether what they were doing in Moneyville was the same or different from what they were doing in school. One girl said, "This is really different from what we do at my school. All we do with numbers is add, subtract, multiply, and divide." This indicated to evaluators that the exhibit provided valuable opportunities for contextualized mathematics learning that children are not experiencing at school.

Not surprisingly, mediation was an important factor in how visitors interacted with the exhibit. Many of the richest math learning experiences we saw involved facilitation of some kind, a sharing of strategies, etc. The deeper math learning in some of the exhibits in Moneyville really emerged when parents or staff facilitated the experiences. In the future, we'd like to look at which characteristics can make exhibits—self-directed, free-choice experiences—as rich as some of the facilitated experiences.

No one visiting that exhibit had any doubt whether or not they were doing math. On the other hand, the Moneyville Exhibit is provocative because, although it engaged a large and diverse group of visitors who enthusiastically participated in activities around money and economics, some of these visitors did not consider their experience to be "mathematical."

Summary and Conclusions

Including mathematics presents new perspectives and difficulties when designing exhibits and programs for families. Much of the wisdom from previous research on science is relevant, but people's attitudes toward math, their avoidance of it, and their limited view of what math is, stretch design and development skills in important ways. We've seen that science centers can help create an environment in which parents can encourage their children's mathematical thinking. As many educators have noted, parental support is vital in promoting children's mathematical development: "Positive behavior toward mathematics must begin in the home. Parents need to promote a love for mathematics in their children—even if they have less than favorable recollections of their own experiences with the subject" (Posamentier, 2006, p. 3). Parents do care a great deal about their children's education and want them to be successful in core subject areas like math. Parents, like teachers, say that raising a child "who wants to learn is the most important role a parent can fulfill" (Farkas et al., 1999,

p. 1). However, they do not always know how they can promote mathematical thinking and enjoyment in their children, other than through the sometimes stressful process of helping with homework.

Science centers can help parents and other family members envision new, more positive roles in the enterprise of informal math learning. If successful, they will provide opportunities for the kind of conversation that began the chapter. The ingredients are deceptively simple: adult, child, a mathematical tool, a question that engages both people, and an environment in which mathematical interactions are encouraged and supported. Science centers can successfully combine these ingredients. It will take thinking creatively about the nature of mathematics, integrating math into the entire planning process, and being able to recognize and support mathematical conversations as they evolve.

References

Barton, A. et al. 2004. Ecologies of parental engagement in urban education. *Educational Researcher* 33:3–12.

Borun, M. et al. 1998. *Family Learning in Museums: The PISEC Perspective.* Washington, DC: Association of Science-Technology Centers.

Bullock, Linda. 2004. *You Can Use A Balance.* Danbury, CT: Children's Press.

Burns, M. (1998) *Math: Facing an American Phobia.* Sausalito, CA: Math Solutions Publications.

Civil, M. 2001. Redefining parental involvement: Parents as learners of mathematics. Paper presented at the NCTM Research Pre-Session. Orlando, FL, April 2001.

Civil, M. and R. Andrade. 2002. Transitions between home and school mathematics: Rays of hope amidst the passing clouds. In G. de Abreu, A. J. Bishop, N. C. Presmeg, Eds. *Transitions Between Contexts of Mathematical Practices.* Dordrecht: Kluwer:149-169.

Coates, Grace Davila and Jean Kerr Stenmark. *Family Math for Young Children.* Berkeley, CA: The Regents of the University of California.

Copley, Juanita V. 2000. *The Young Child and Mathematics.* Washington, DC: National Association for the Education of Young Children.

Farkas, S. et al. 1999. *Playing their Parts: Parents and Teachers Talk about Parent Involvement in Public Schools.* New York: Public Agenda.

Faulkner, Keith. *So Big! My First Measuring Book.* Los Angeles, CA: Little Simon.

Hartog, M. and P. Brosnan. 1994. Doing mathematics with your child.

ERIC/CSMEE Digest. Columbus, OH: ERIC/CSMEE (updated June 2003).

Kliman, M. *et al.* 2006. Math out of school: Families math game playing at home. *School Community Journal.*

Kralovec, E. and J. Buell. 2000. *The End of Homework.* Boston, MA: Beacon Press.

Korn, R., (1996). *PISEC baseline visitor study.* Unpublished manuscript.

Lionni, Leo. 1995. *Inch.* NY, NY: Harper Trophy.

Lubienski, S. 2003. Is our teaching measuring up? Race-, SES-, and gender-related gaps in measurement achievement. In Clements and Bright. Eds. *Learning and Teaching Measurement: 2003 Yearbook.* Reston, VA: National Council of Teachers of Mathematics, Inc.:282–292.

Pluckrose, Henry Arthur. 1994. *Math Counts: Weight.* New York, NY: Watts Books.

Posamentier, A. 2006. Add math at home where it counts in kids' lives. April 14, 2006, edition of Newsday.com.

Randi Korn and Associates. 2005. *Fort Worth Museum of Science and History, Remedial Evaluation, Lone Star Dinosaurs.* Alexandria, VA: Randi Korn and Associates.

Scherer, Marge. 1998. Engaging parents and the community in schools. *Educational Leadership* 55(8):5.

Schliemann, A. D. and D. W. Carraher. 1992. Proportional reasoning in and out of school. In *Context and Cognition.* P. Light and G. Butterworth. Eds. Hemel Hempstead: Harvester-Wheatsheaf:47–73.

Thompson, V. <www.enc.org/focus/family/document.shtm?input=FOC-000729-index> (June 23,2005).

CHAPTEr 7

Connecting Schools and Science Centers Through Mathematics

Tracey Wright

What is the role of science centers in supporting mathematics literacy by enriching and complementing what's happening in schools? Currently, school systems are connected to science centers through the obvious disciplinary links with science. However, science centers can also be a key resource for schools when it comes to math learning. In working with schools, science centers bring to the table key ideas:

✓ Everyone learns math in different ways.

✓ Children need experiences that engage them emotionally, physically, and intellectually.

✓ Significant math ideas take time to build.

✓ It's important to preserve both the individual and interpersonal nature of learning.

These ideas, rooted in Dewey's (1938) philosophy and expanded upon through various models of "constructivist learning" (see Hein and Alexander, 1998), are critical to math learning in science centers, as seen throughout this book.

Constructivist models, which stress the idea that the learner is an active constructor of ideas, are also the basis for a significant amount of mathematical learning in schools. National reports emphasize that the mission of school math programs is to develop deep understanding of the discipline, stating, "All young Americans must learn to think mathematically, and they must think mathematically to learn" (National Academy of Sciences, 2001, p. 16). Newer math curricula, developed with funding from the National Science Foundation and used in approximately 25–40% of schools nationwide (Bement, 2006), underscore the central role of the learner in solving mathematical problems where there are multiple entry points, multiple solution paths, and sometimes even multiple answers. When students and teachers who use these curricula visit science centers, they often appreciate opportunities to explore in more depth mathematical questions that arise in their curricula.

Of course, there are many classrooms and schools that employ more traditional approaches to learning, where mathematics is seen primarily as a body of knowledge and processes to be mastered incrementally. While designing programs for schools, it may be tempting to emphasize all of the mathematical "bits of knowledge" that students might encounter on their visits to the science center. However, we have found in our work with science centers that even schools that employ more traditional approaches to teaching and learning value active math learning that places students at center stage and also value the importance of math while doing science. Of course, many educators also want to see that the math done in science centers is linked with "school math," particularly with state and national math standards.

Sometimes, because schools are under pressure to increase students' scores on high-stakes tests, it is tempting for science centers to market their school programs by proclaiming how well they link with the content of these tests. And if science centers are focusing on math content emphasized in curriculum standards, it's likely that they are in a general sense helping students become familiar with mathematics that the students are expected to master. However, it is extremely unlikely that science-center programs, or even short-term school interventions themselves, make a critical difference in test scores. When working with schools around the topic of math, "It's terribly important that you make no false promises and only do what you can. You need to remind yourself and tell your clients, the schoolteachers and administrators, that *science museum education programs will not improve test scores*" (Hein, 2001, p. 12, emphasis in original).

Yet there are many important outcomes, other than test scores, for which science centers can play a key role in fostering students' mathematical achievement. In considering the wide variety of school math programs that exist in most communities, it makes sense for science centers to focus on the unique mathematical strengths that they can offer to school visitors. Once these mathematical strengths or specialties have been identified, they should be analyzed to determine how they could supplement, enhance, and expand upon what is happening in schools. Although this process of articulating mathematical strengths will be unique to each science center, there are at least three general ideas for centers to keep in mind as they work with schools on math learning. These are described below and followed by three examples from centers that have developed approaches that complement the needs of schools.

1. Science centers offer schools a context and purpose for bringing math and science together: In formal education, interdisciplinary learning—especially learning involving math and science—is recognized as critical to creating literacy and to generating new knowledge. Educators across the board recognize that there is no science without mathematics. Yet the departmental structure of schools, as well as the tightly defined content of standardized tests, often precludes making important, natural connections between math and science. Fortunately, a key strength of science centers is that they transcend disciplinary boundaries. Science centers show how different topic areas are collectively brought to bear on a problem and how tools and technologies are used, as well as ways in which evidence is considered and interpreted.

2. Science centers can provide an expanded perspective on mathematical topics: This is especially true with respect to math that involves data, measurement, patterns, algebra, and calculus. Schools are seeking to broaden their perspectives on math learning. Both national and state math standards encourage learning that goes substantially "beyond arithmetic." These same standards encourage students to pose their own questions, find different ways of solving problems, and critically evaluate their solutions. Science centers shine at showing processes such as questioning, using evidence, and making interpretations. In short, science centers can support schools in broadening students' ideas about what math is in the same way that they support schools in broadening students' views on the nature, process, and content of science.

3. Science centers have a vital role to play in engaging all students in math by making it come alive: Recognizing that many students lose interest in math during middle school, schools are striving to bring more and more students into the mathematical pipeline. Students need to be able to engage in much more challenging mathematics than is typically taught in the American K–12 system. As pointed out in earlier chapters, this is an important social justice issue because a disproportionate number of students of color and poor students drop out of math and are subsequently excluded from pursuing many higher-paying jobs and scientific careers. Because science centers make a point of encouraging different learning styles, such as kinesthetic learning, experimentation, and collaboration, they can help schools reach more students.

In the descriptions that follow, we show how an aquarium and two science centers have built mathematical connections with schools by addressing schools' concerns with interdisciplinary learning, expanding students' notions of math, and bringing all students into the mathematical pipeline.

Example 1: New England Aquarium's Penguin Program

The New England Aquarium (NEAq) in Boston, Massachusetts, has developed an interesting mathematical experience for young children that involves the study of penguins. Although it is offered to grades K–2 and 3–5, it is used primarily by the first and second grades. This data-enriched experience based on penguin behavior was designed as an interactive alternative to a fact-based penguin presentation in which information was presented to a seated audience. During the interactive program, students collect and analyze data on penguin behaviors. But simply adding math on to an existing program was not the main goal. The NEAq wanted to create a more engaging program; using data was a useful tool to achieve that goal.[1]

The enhanced 45-minute program began with an optional pre-visit experience in which teachers helped children make identification bracelets similar to those worn by the penguins at the aquarium. Once at the aquarium, children were involved directly in collecting data on seven penguin behaviors (swimming, bowing, trumpeting, preening, pointing, resting, and walking, which the children learn through role-playing. After enacting these behaviors, children were given observation charts and told how scientists might collect data by noting the behaviors they see at regular time intervals.

At this point, half the children were assigned the role of penguins and half became "scientists" who each observed and charted the behavior of a particular simulated penguin. After a few minutes, the groups switched, enabling everyone to have a turn role-playing a scientist as well as simulating a penguin.

To analyze their data, each member of the group counted the number of times he or she had observed each penguin

[1] Neither the old nor the new program uses live animals, a common expectation of visitors.

MATH MOMENTUM

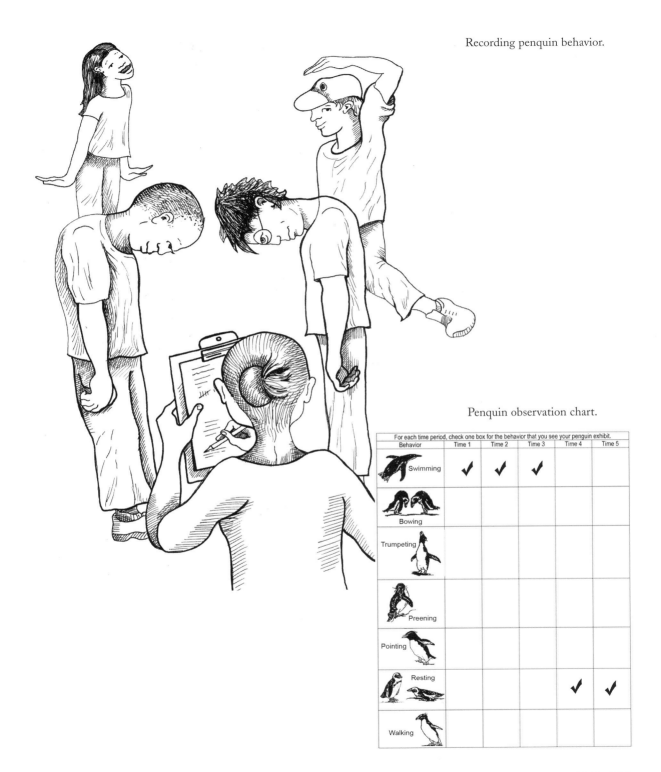

Recording penquin behavior.

Penquin observation chart.

For each time period, check one box for the behavior that you see your penguin exhibit.					
Behavior	Time 1	Time 2	Time 3	Time 4	Time 5
Swimming	✔	✔	✔		
Bowing					
Trumpeting					
Preening					
Pointing					
Resting				✔	✔
Walking					

Voice from the Field

Enacting Penguin Behavior

Rebekah Stendahl, Supervisor of Family Programs, New England Aquarium

At the New England Aquarium, we welcomed the opportunity to redevelop one of our programs by incorporating math concepts. Those of us who were working on the math team had past experience with both informal math education and incorporating real science skills in programming for children. We knew that this could be a means to a more engaging program. Because we knew early on we would be focusing our project on data, we wanted to use a live animal collection so we would have a strong data set that visitors could access and, ideally, collect for themselves. By focusing on data, we knew we could make this program more engaging, even without the excitement of a live animal.

Our main hope in incorporating data was to find a way to provide a more memorable experience for our program participants. All of our programs have an underlying conservation message that is appropriate to the age of the audience and the topic of the program. If we hope to drive this message home, we need to find a way to engage the participants in the subject matter and make the experience more meaningful. In this way, we can work to achieve the NEAq's mission to "present, promote and protect the world of water." With most interactions ranging in duration from less than a minute to an hour, we can't expect to achieve lasting change by simply sharing basic facts. But by engaging our visitors with dynamic programs, we can hope to instill a desire to protect for years to come.

After working with the penguin husbandry department to find a suitable topic for data collection, we began to develop a program that would focus on collecting data on penguin behavior in order to learn more about the colony as a whole. However, because we aren't able to bring even one penguin out on programs, much less the entire colony, we would need an alternative that would allow students in classrooms to collect real data. At this point, reminded of the popularity of role-play activities in the early elementary grades, we realized that although the data collection had to be a real experience, the source of the data could be pretend. We developed a program where children acted like penguins. At the end of the program, after we have analyzed the collected data, educators discuss with children how realistic the data can be. There is always an understanding among the children participating that they would most likely see differences between their data and actual penguins if they observed a real colony of penguins at the aquarium. They may not be able to guess what those differences would be, but they are curious to find out.

There is a lot of interest in the work of the

penguin husbandry staff when they are observed in the exhibit. Participants in our penguin class have wondered whether the data sheet we use in the program is the same sheet they have seen the husbandry staff using on their clipboards in the exhibit. In fact, the data that the husbandry staff collects is related most often to feeding. The data sheet we use to observe behaviors in the class is something that we created especially for the school program. This is one example of how the program was developed to be successful for the target age group and not tied too exactly to what is happening behind the scenes at the aquarium.

The class was originally developed with a third-grade target, based on the skill set needed for the data collection activity and how it related to state and national standards. However, we quickly discovered that most of our school interest was coming from the first and second grades, primarily in January and February, when these classes are studying penguins in relation to a cold-weather curriculum. We found ourselves tweaking the details of how we deliver the program, but we were able to keep the overall structure and message.

The act of data collection for most of these classes became a first exposure for children to look at a table with rows and columns and learn what it means to make observations. The 30-second time intervals between observations began to vary as we waited to make sure all the students were ready to proceed before moving onto the next observation time. Some children we watched seemed to be checking off the behaviors they wanted to see, rather than the ones their partners were actually exhibiting. Quite often the younger children would be creating data sheets where almost every box was checked. But upon looking closely, we saw that all these children were still engaged in the program, even if they weren't clear on the data collection method. By making sure we always took time at the end of the class to analyze some data (rather than hoping the teacher would be able to take this step upon returning to the classroom), we could help bring the kids who didn't quite understand the process back into the program with an overview of the bigger picture.

Ultimately, the overarching message of this program is that you can learn more about the whole by taking time to study the parts. As program developers, our hope is that we have created a program where children can see themselves as scientists and as part of the scientific process. With time and experience, they will perfect the skills we may have first introduced with our programs.

behavior. Three of these behaviors became the focus of data analysis. Each child selected the number of stick-on notes that corresponded with the number of times he or she had observed the three behaviors. Then the children built a cumulative graph of their data with stick-on notes, showing the total number of times these behaviors were observed.

As children examined the graph, they were asked to note which behaviors occurred most and least frequently, and to compare the numbers in each category. Then they were asked questions such as, "Your penguin was resting all the time. Is that unusual?" Or, "How do you think the data from real penguins might look?" In addition to exploring questions about penguins, this visual experience also helped students make the connection to previous school learning about graphing.

NEAq observed that children who participated in the data-enriched penguin program were highly engaged in the program in a way that was qualitatively different from that of children who had participated in the traditional fact-based penguin program. Students were moving around, enacting penguin behaviors, and physically moving around stick-on notes to create a graph. Not only were students motivated during the activity, but also, during discussions, children talked about wanting to collect additional penguin data themselves. Staff also noticed that students who visited the live penguin exhibit after their program were spending more time at the exhibit itself. They further hypothesized, based on these findings, that children in the new program were more keenly observing actual penguin behavior and were more able to identify the behaviors they had enacted themselves. One staff member commented specifically on the importance of focused observation: "Often at the aquarium, children (and adults) bounce from one exhibit to another, finding it hard to become really engaged with one exhibit or animal because there is so much more to see. Visitors could stand for a long time and look around at the penguins; this activity encourages people to slow down and really study one penguin, which can be a more beneficial or meaningful way to observe animals." Including visual and tactile as well as auditory experiences is something that museum staff may already do in science, but may not see how to do as easily in mathematics.

In this example, the NEAq is taking an interdisciplinary approach to learning math and science by demonstrating how

Graphing penguin behavior.

scientists might use data to learn more about animal behavior. This is a way for students to see the meaning of the math in the context of doing science. By collecting data to answer questions, students are opening some of the black boxes of science. Even if their data sheets are not always right or clear, they are getting a glimpse of why charts are used as a system for collecting data. The experience of collecting and analyzing data has become a tangible way to justify and answer a question important in both science and math: "How do you know?" This illustrates an important purpose in looking at data representations to understand a broader phenomenon (in this case, penguin behavior), thus tying together math and science in conceptually relevant ways.

How does this work complement and support what might be happening with data literacy at school? One significant contribution is by showing how subject areas actually interact, rather than existing as isolated disciplines. Although it is not impossible, it is often more difficult to integrate math and science in schools. Attempts at sustaining integrated curricula in schools have faltered, often because of pressures to cover mathematical or scientific material in a certain manner. Sometimes material in school is presented to resemble that which students will face on a standardized test. Often, tests are required in math, but not in science, which may lead teachers to emphasize math and de-emphasize science.

Another way in which the penguin activity supports what happens in school is by enriching the context for the study of data. Although the National Council of Teachers of Mathematics calls for a greater emphasis on data (NCTM, 2000), collecting data for a purpose is difficult to do at school. Because of limited time, students may do simple surveys about themselves (favorite ice cream flavor, number of letters in their names, etc.), but they don't often have the opportunity to focus on collecting and analyzing data that have a real scientific purpose.

There are other reasons besides the time factor why managing data investigations in schools can be problematic. Although students and teachers often get excited about this kind of work, there's often too much walking around or too high a noise level. On the other hand, these activities are very appropriate for science centers, and teachers appreciate not having the noise and mess in the classroom.

A final important feature of working with data in the context of animal behavior is that it can change the ways in which parents think about math teaching and learning. In the penguin activity, the role of parent chaperones evolved. Parents helped kids chart their data, rather than simply monitoring student behavior. This example illustrates how incorporating math into a science-center program can broaden students', teachers', and parents' ideas about math and help them see math as alive and engaging.

To help schools explicitly link the penguin activity with important mathematical goals, the NEAq designed a brochure on school programs that correlates the penguin program with goals of the Massachusetts Curriculum Frameworks (2000) in mathematics (as well as in science and technology). To make the link even stronger, staff emphasize how the activity enhances the work that children from the Boston city schools are doing with data as they work with a new mathematics curriculum. Linking to the curriculum and standards used by key constituents helps teachers explain the value of field trips to administrators and parents.

Example 2. Mathematical Questions & Things to Do at the Museum of Science, Boston

Graphing at the Human Body exhibit.

To support the current efforts in mathematics education, science-center educators need to know what school mathematics looks like in its best form. Though there are many ways to learn about this (such as visiting schools, videotapes, books, etc.), the extensive collection of NCTM resources is an important source of information in this endeavor (see www.nctm.org). Science-center educators may be surprised at how familiar many of the Principles and Standards already are. In fact, the NCTM process standards and many aspects of the content standards complement the goals of science education as articulated by both the American Association for the Advancement of Science science standards (AAAS, 1993) and the National Science Education Standards (NSES, 1996). For example, the importance of representations, problem-solving, communication, and doing math in context are all critical elements of NCTM's vision. One standard, explicitly called "Connections," encourages teachers to enhance mathematical understanding by drawing from informal, everyday experiences, as well as other disciplines (especially science and social studies), as a source of problems.

Working with the understanding that the NCTM standards are a good starting point for science centers to connect with school mathematics, the Museum of Science, Boston, mapped these standards with four exhibits. They produced a resource

Human Body Exhibit

Mathematical Questions & Things to Do	NCTM Standards
• Are the very tallest (longest) babies boys or girls? • Where should a seven-year-old stand to be measured? • During what years is the range of heights for girls greater than the range for boys? • Can you determine a typical height (mean or median) from the information on this graph? • When does the female growth spurt occur? When does the male growth spurt occur? • What is your height in inches? In centimeters? • What is the range of heights in your class?	**Measurement** Apply appropriate techniques, tools, and formulas to determine measurements. Pre-K–2: Use tools to measure. Pre-K–2: Develop common referents for measures to make comparisons and estimates. 3–5: Select and use benchmarks to estimate measurements. **Data Analysis** Formulate questions that can be addressed with data, and collect, organize, and display relevant data to answer them. Pre-K–2: Pose questions and gather data about themselves and their surroundings. Select and use appropriate statistical methods to analyze data. Pre-K–2: Describe parts of the data and the set of data as a whole to determine what the data show. 3–5: Describe the shape and important features of a set of data, and compare related data sets, with an emphasis on how the data are distributed.

called Mathematical Questions & Things to Do that helps teachers and students draw out the math at these four exhibits. Shown here is a sample of this work from the Human Body Exhibit, drawn from a larger, searchable database developed by the musum Educator Resource Center (or see www.mos.org/doc/1812?id=578).[2]

Developing these connections with standards was important to museum staff because the museum's mission, "Stimulating interest in and furthering the understanding of science and technology," means that staff are hoping that learners will recognize math as a valuable discipline not only in and of itself, but also as a critical tool for understanding science and technology. Linking exhibits to the NCTM standards is a way of making explicit what children are working on, why it is mathematical, and how it relates to school math. This work goes beyond simple labeling, by describing the mathematical ideas in an exhibit and how these can be probed more deeply. Teachers can then customize a field trip to focus on a mathematical idea or on a particular set of exhibits that relate to their curricular goals.

[2] The Educator Resource Center includes standards-based science, technology and engineering resources such as Museum exhibits and programs, books, video, curriculum, and others educational resources. The Museum currently maps its resources to the National Science Education Standards, the Massachusetts Science and Technology/Engineering Curriculum Frameworks, ITEA Standards for Technological Literacy, and the National Council of Teachers of Mathematics Standards. See www.mos.org/doc/1369.

Voice from the Field

Where's the Math?

Loren Stolow, Camp-Ins Program Coordinator, Museum of Science, Boston

One of our first steps to increasing math awareness and education at the Museum of Science, Boston, was to gather staff with diverse perspectives and ask ourselves, "Where is the math that already exists in the Museum of Science exhibits and programs?" This process proved valuable for several reasons. We were able to bring together people from different departments who do not regularly work together; we were able to strengthen staff's understanding and awareness of math in the museum and math education; and we were able to better design math resources for visitors.

Gathering People Together

Our first meeting was a brainstorming session, open to anyone. It was attended by 20 people from many parts of the organization—including exhibit content development, public programs, current science, discovery spaces, school programs, collections, and curriculum development—all of whom had varied expertise and perspectives. People's backgrounds and comfort with math varied as well, from math "geeks" and former math teachers to those who didn't really like math very much. As the conversation progressed, we developed an extensive list of existing math-related exhibits and programs. There were obvious places to find math, such as our Mathematica exhibit.

But we also talked about the concepts of size and scale that are highlighted in the Dinosaur Exhibit; graphs in the Models Exhibit; activities that develop number sense in Computer Place; Camp-In programs about patterns and probability; teens involved in data collection; curriculum development projects that link science, math, and engineering; and many other places we find math in the museum.

This process was important in shaping our future conversations and work. Once we identified the math in the museum for ourselves, it seemed clear that we needed to do the same thing for our visitors. We needed to make the math more accessible and, particularly for classroom teachers, connect the math in the museum to the concepts and standards they use in their classrooms.

Professional Development

In our conversations we talked about what mathematical thinking looks like in the museum, drawing on our own understanding of math and the national math education standards. It was important for everyone to look beyond numbers and broadly define mathematical thinking. We found numerous opportunities for visitors to collect, manipulate, and analyze data; to make measurements; to classify and sort materials; and to solve problems using math. We observed visitors in the museum exhibit halls to discover how they were using exhibits mathematically.

We hosted a workshop, Math in Museums: Using Data Effectively, attended by staff from informal science institutions throughout New England. Many Museum of Science staff also attended the workshop. After learning about and practicing concepts of data collection and analysis, participants spent time observing visitors in several of the Museum of Science exhibits that were potentially mathematically rich. Although we may work full-time in a museum, we don't always give ourselves permission to take the time to step back and observe. Participants were able to describe behaviors that demonstrated mathematical thinking—counting, finding patterns, and changing variables. They were also able to identify opportunities for mathematical thinking, even if it were not yet happening.

Strengthening our own understanding and familiarity with mathematical thinking and how it happens in museums is essential to improving our ability to promote mathematical experiences for visitors.

Designing Resources

Our brainstorming sessions led us to create online resources that described ways for anyone to do math at the Museum of Science. This connected well with an ongoing project to create a Web-based educator resource center that includes a wide range of materials. Subsequent interdepartmental meetings focused on identifying the math in particular exhibits. We continued to draw on experience and expertise from different parts of the museum. The meeting about the Human Body Exhibit, for example, included several exhibit developers, a school program presenter, an evaluator, the Camp-In coordinator, a Discovery Space volunteer, and a math curriculum developer. Together we identified components that had strong math connections and developed charts of Mathematical Questions & Things to Do, along with pictures and exhibit descriptions. We then identified math education standards that related to the concepts presented in the exhibits. So far we have developed resources to accompany four of our permanent exhibits. There is still much more to be done. A small group of middle-school teachers reviewed early drafts of the questions and responded positively, but we do not yet know how teachers are using these online resources and incorporating them into field trips with students. We have many ideas about how to improve and expand the resources, such as adding more exhibits, separating information by grade level, and adding information about general math concepts.

Our experience demonstrates that mathematical ideas are widespread in science-center exhibits and programs. By working with people throughout the institution, there are ways to strengthen our own awareness of the math that exists and make it possible for students and teachers to stretch their understanding and have richer mathematical experiences.

In this "Voice from the Field," we see how using the NCTM Standards is a way of connecting to schools and to mathematical ideas. The museum is making explicit some of the deeper connections between certain science exhibits and specific math standards by posing questions to investigate. According to science-center evaluator Cecilia Garibay, "It's not enough to know what the Standards are; one must be able to interpret them to museum staff and sometimes even to school staff" (Garibay, 2002). By going beyond a simple correlation in which math standards are used to justify a field trip, the Museum of Science, Boston, is offering educators the opportunity to develop engaging and connected learning experiences in math and science.[3]

In addition to using math standards as a way to understand science more deeply, the Boston Museum of Science's work also illustrates the way in which NCTM Standards are used as a resource for understanding what mathematics is. Museum staff approached math standards in the same spirit in which they approach science standards, recognizing that they are not as daunting or as narrow as some may think. They started from a real problem and avoided overstructuring it. The process described by Loren Stolow helps us see the ways in which learning math for yourself, doing math with colleagues, and examining how visitors are learning math concepts are crucial steps in supporting school groups as they learn mathematics in the context of science.

Example 3. Science Museum of Minnesota's Math Packs Field Trips

In a time when schools are under scrutiny and expected to make every instructional moment count, teachers increasingly want to know, "Why should we visit science centers? What will my students learn and how will it make a difference in their academic success?" As seen in the previous example, it is important to examine both science *and math* standards as a starting place for school programs. Teachers report that connections with standards help them to justify a field trip. One teacher puts it this way: "At least I know this program is not just some wacky tangent!" But the question remains: How do teachers and students benefit mathematically from field trips to a science center?

This question is addressed by the Science Museum of Minnesota, which has developed a school program, MathPacks: Measuring Growth, for fifth-graders. Basically, the program involves measuring dinosaur femurs and tree rings to

[3] In addition to supporting mathematical engagement for school groups by reaching out to teachers, "The Museum of Science is dedicated to providing a wealth of resources to educators of all kinds. Formal educators, informal educators, homeschoolers and anyone else who needs materials to enhance and accompany their teaching..." (http://nmhm.washingtondc.museum/news/links2.html)

understand growth (see www.smm.org/mathpacks). They developed this program in order to provide more math opportunities for school groups visiting the museum. The program consists of 60 backpacks that contain measuring tools such as rulers, tape measures, and calipers. MathPacks also contain museum gallery maps, a handheld lens, clipboard, calculator, pencils, and a juvenile dinosaur femur to be used during the field trip. Also included are name tags for students with preassigned group roles such as Measuring Master, Equipment Handler, Recorder, and Location Expert/Navigator. Small groups of students share one MathPack.

Teachers are given pre-visit activities to do with students to introduce the topic and skills, and post-visit activities that allow students to share some data and their analysis. To introduce the program at school, an online video features their Curator of Paleontology, Kristi Curry Rogers, who describes specific ways that paleontologists use math in their work. Dr. Rogers asks for kids' help collecting data about dinosaur growth.

Measuring the apatosourus femur.

Following is a "Voice from the Field" describing how this program was developed and how it connects to school math.

Ten fifth-grade teachers (representing eight schools and

Voice from the Field

Seeing Math in Action

Chris Robinson, Program Developer, Science Museum of Minnesota

I would say that the most compelling standards-and-curriculum connection is with the Apatosaurus femur work we are doing with measurement. We looked directly at the fifth-grade Minnesota Standards in Mathematics *(2003), and we also looked quite a bit at the curricula being used in the area. Many of our participants used a new mathematics curriculum, where children are actively engaged in problem-solving. After a discussion with our paleontologist Kristi Curry Rogers about measuring growth in Apatosauri, and knowing that we had an adult Apatosaurus femur on the floor accessible to measurement, I looked at the curriculum and found that there were a lot of activities focused on ratios. We asked Kristi whether perhaps Collections had a young Apatosaurus femur that we could use to make casts. They did and so we developed an activity where students compare measurements of the adult and baby femurs (femora) in a variety of ways to encourage a better understanding of ratio. For example, they not only calculate how many times bigger (or longer) the adult is compared to the baby, but they are also asked to figure this out another way, for example, by using string or the femurs themselves.* The Minnesota Standards in Mathematics *that we felt were addressed strongly were these:*

II. Number Sense, Computation, and Operations
B. Computation and Operation
Compute fluently and make reasonable estimates with fractions, decimals, and whole numbers, in real-world and mathematical problems. Understand the meanings of arithmetic operations and how they relate to one another.

III. Patterns, Functions, and Algebra
B. Algebra (Algebraic Thinking)
Represent mathematical relationships using equations. Evaluate numeric expressions in real-world and mathematical problems.

V. Spatial Sense, Geometry, and Measurement
C. Measurement
Measure and calculate length, area, and capacity, using appropriate tools and units to solve real-world and mathematical problems.
Select and apply the appropriate units and tools to measure perimeter, area, and capacity.

over 400 students) participated in pilot testing the MathPacks program. Six were suburban public schools, including two magnet schools, and two were parochial schools in partnership with the museum. To find out more about how teachers responded to the program and what students were learning, Science Museum of Minnesota staff videotaped all the field trips and afterwards interviewed (and videotaped) all ten teachers. Several student interviews were recorded as well.

Results showed that many of the teachers participated in this program because they were enthusiastic about mathematics. Having a math field trip was exciting and unusual. In a few schools, the teachers team teach. One teaches math and science; the other, social studies and history. These teachers in particular commented on how great it was to integrate their math and science work. One teacher pointed out the differences between this field trip program and a more open field trip. "This field trip had a focus. It was hands-on and had tools to use at the museum. Kids were forced to slow down. Even with a hands-on exhibit, the kids don't always wait to engage in depth—they push buttons, see what happens, and then move on."

Activities with a focus often have clear links with school math. For example, one of the MathPacks activities asked students to measure a femur in a dinosaur mount about 1.5 meters above the floor, surrounded by an exhibit fence. The activity had to do with measuring a dinosaur femur at two different ages to get a sense of the size of the whole animal, by using the idea of similar triangles, a technique often taught in isolation at school. By using the idea of similar triangles, visitors could estimate the length of this object that they could not measure directly. In a pre-visit experience, students had familiarized themselves with these skills by measuring a flagpole indirectly, making use of similar triangles. Teachers reported that this activity was hard, but that students really liked the challenge of similar triangles. After the field trip, one student reported, "We studied similar triangles, but now we know why we use them." Using this mathematical technique was a useful way to collect physically inaccessible data in order to consider the question at hand. And in fact, the way large dinosaurs looked, walked, grew, and moved is a topic still being explored today by scientists like Kristi Curry Rogers. So in the learning situation described, students are not only making connections between math and science, but between ideas studied in school and those in the real world.

According to Maija Sedzielarz, School Visits Program

Coordinator, teachers had many positive things to say about how students benefited from the program: "Many teachers think this program is successful mostly because kids have a great time and are engaged." Teachers talked about how they really liked real-world applications. "Hands-on experiences matter. They're not something the kids necessarily get at home. And they can't just read about how big a dinosaur is or even see that in a movie. They have to go and stand next to a dinosaur to get a sense of how big it really is." Teachers from a parochial school reported that their students (who all visit a few times a year, every year) have become more independent learners. One teacher pointed out that her kids don't see math as hard in this setting. Another teacher whom Maija spoke with before the field trip said she had been concerned initially that students who perform poorly in math would let other, stronger students do all the work. However, after the experience, she remarked that all students were involved in the work. "All the kids could be successful; they all had a role."

Teachers were also asked whether students referred to their MathPacks experiences after the trip. This happened in a few ways. One teacher described students' recalling their museum experience while in math or science classes: "When we did our variables unit in science, with length changes, some kids remembered using measuring tapes, a familiar tool—and also when we studied ratios." Another teacher said, "When asked about the top three events of the year, this field trip was often mentioned." In addition, a group of kids who went back to the science center for a different field trip asked whether they could use the MathPacks backpacks again. So although we cannot always predict how an experience such as this may impact learning, it is clear that for many students, these field trips were engaging, relevant to math and science, and memorable.

Not only was the field trip was valuable to students, it also gave teachers a chance to develop new perspectives on students, as well as on the topic being studied. Many teachers want to know more about how their students learn, but they don't have enough opportunities in school to step back and examine student learning. Recall the teacher who had concerns about poorly performing math students letting other students do the MathPacks work. She observed, however, that this did not happen and gained insight about what is needed to involve all children in a math activity. Pat Carini (2000) points out that "when a teacher can see… the child engaged in activities meaningful to her, then it is possible for the teacher to gain the insights needed to adjust her his or own approaches to the child accordingly." Even one engaging experience can shape the way teachers see students, the way students view themselves, and the possibilities for future learning experiences.

Conclusions

In this chapter, we have described a variety of ways in which science centers are reaching out to schools by focusing on math learning. In connecting with schools around the teaching and learning of mathematics, science centers work towards their mission of demystifying science, as well as expanding visitors' notions of the centrality of math to scientific work. They also use math to build upon unique aspects of their mission, as was the case for New England Aquarium with its linkage of math to its broader conservation message. Beginning with existing strengths is critical. The Science Museum of Minnesota chose to incorporate math with paleontology because of the strong connections they already had in this field and with a paleontologist on staff. It should be noted that both of these programs involve a commitment to staff resources, which is often not as great a funding priority in museum settings as exhibit development.

In all of the work discussed in this chapter, strong partnerships with teachers and schools were essential from the earliest stages of development. Teachers reviewed science-center math activities and programs during the design phases and early phases of implementation. Teachers often served as advisors to specific projects. As Andrea Anderson points out in her pioneering work on math in science centers, "Exhibits, programs, and materials are more successful and effective when relationships are built with end users as partners who are equal contributors to the outcomes" (Anderson, 2001, p. 105). Anderson's work showed that strong partnerships between science centers and schools enabled science centers to better meet the needs of their audiences and to find the strongest and most appropriate roles to play with respect to schools. These findings suggest that to promote math learning, centers should focus equally on their partnerships with schools and on their own unique missions and goals.

Through the three examples, we have seen ways in which science centers approach math in the same way they approach science: through a process of inquiry. We've seen students dealing with math in a way that isn't isolated or dry; it's scientific and engaging. We have seen students approaching ideas in ways that deepen their mathematical understanding and build their understanding of science. But although we've seen a range of students invited into math through engaging and relevant programs, what we have yet to see is more thorough research on mathematics learning in informal settings.

Bartels and Hein (2003) remind us too that one body of work largely absent from the field of museum education is action research, or research conducted by practitioners themselves. Both schools and science centers have important roles to play in determining the impact of science-center math programs on students' learning and engagement in math.

References

American Association for the Advancement of Science, Project 2061. 1993. *Benchmarks for Science Literacy.* Oxford University Press, NY.

Anderson, A. 2001. *Mathematics in Science Centers.* Washington, DC: American Association of Science-Technology Centers.

Bartels, D., and G. Hein. 2003. Book review: Learning in settings other than schools. *Educational Researcher.* August/September: 38–43.

Bement, A. 2006. Testimony before the Committee on Science, U.S. House of Representatives hearing on K–12 science and math education across federal agencies. March 30. Washington, DC.

Carini, Patricia. 2000. *From Another Angle.* NY: Columbia University.

Dewey, J. 1938. *Experience and Education.* NY: MacMillan.

Garribay, Cecilia. 2002. Museum-school partnerships: Lessons from the field. *Current Trends in Audience Research* 15:73–83. American Association of Museums.

Hein, G. and M. Alexander. 1998. *Museums: Places of Learning.* Washington, DC: American Association of Museums.

Hein, G. 2001. High stakes tests don't belong in science museums: We can do better than that! Paper presented at Association of Science-Technology Centers annual meeting. Phoenix, AZ. October 10, 2001.

Massachusetts Curriculum Frameworks. 2000. Malden, MA: Massachusetts Department of Education.

Minnesota Standards in Mathematics. 2003. Roseville, MN: Minnesota Department of Education.

National Science Education Standards. 1996. Washington, DC: National Academy Press.

National Council of Teachers of Mathematics, Inc. 2000. *Principles and Standards for School Mathematics.* Reston, VA: NCTM.

CHAPTER 8

Feeling Math Confident Outside of School: Math in Youth and Outreach Programs

Jamie Bell

They were working kinesthetically with the concepts of volume and density, and the kids didn't leave feeling stupid or dumb. It heightened our expectations of what the Lawrence Hall of Science has to offer our community, and we hope to work in similar ways with them in the future.

Shirley Brower, Director, South Berkeley YMCA

Ms. Brower's heightened expectations are attributable to the new ways that the Lawrence Hall of Science found to incorporate their work with math into both their ongoing teen program and an outreach program in the community. Mathematical outreach is a logical next step for centers that are already doing science-based outreach. In most cases, this work involves incorporating math into existing programs, rather than starting new programs from scratch.

Within science centers, there are a rich variety of on-site explainer programs, camps, after-school clubs, classes, and long-term programs for teenagers, many of which provide educational enrichment as well as volunteer and employment experiences for their participants. Many centers also offer off-site programs where museum staff bring portable exhibits and activities by van or car to community centers, libraries, schools, or hospitals. The target audiences for both these types of programs are often school-age youth, whose exposure to informal science experiences enhances and complements what they are learning in the classroom. In doing youth and outreach programs, science centers provide their partner organizations with tools they can use to add value to their programs and further their mission of combining learning experiences with fun. In turn, schools and community-based organizations provide partnerships with a broad range of audiences to science centers, often with the opportunity to work with these audiences over an extended period of time.

Issues

Several major questions must be addressed in making math outreach programs engaging and relevant to young people. All are based on the need for elementary through high-school youth to have meaningful educational experiences that have a clear purpose. The questions "Why are we learning this?" and "When are we ever going to have to use this?" that students often ask about school math are an important starting point for science centers in designing ways of incorporating math into their youth and outreach programs. More specifically, when building math-rich youth and outreach programs, science centers should begin by asking themselves these questions:

✓ How does this program address young people's need for meaningful work, including service and community work? (middle- and high-school youth)

✓ How does this program impact young people's motivation to do well in school? (youth of all ages)

✓ How does this program meet young people's need to make math relevant, engaging, and fun? (youth of all ages)

Increasing youth's confidence and engagement in math is vital, and science centers can play a crucial role in this enterprise. As mentioned in previous chapters, many students become disenchanted with math by middle school and may become math avoidant. The ingredients associated with math avoidance include spending time in classrooms where getting the right answer is emphasized, not much discussion takes place, and there is public scrutiny and performance pressure. On the other hand, math avoidance is lower in classrooms where there is more humor, more modeling of questioning by teachers, and more peer support and collaboration. (Turner, et al. 2002). Fortunately, these qualities are associated with many science-center outreach programs.

Working on increasing confidence and engagement in math is a good beginning for youth and outreach programs, but it is not enough. These programs need to be grounded in rigorous, relevant mathematical content. Youth and outreach programs are particularly well-suited to engaging young people in nonroutine problem-solving that involves mathematics in the context of everyday life.

Using mathematics in the service of everyday problem-solving is a critical skill, and one that U.S. students lack. Recent international studies show that American 15-year-olds lag far behind their peers in other industrialized nations with respect to problem-solving that involves math (PISA, 2003). The problems on this major international assessment were not particularly difficult from a mathematical standpoint, though they involved a few steps. For example, one problem involved planning the best route for a vacation. Students were first asked to use a map and table to calculate the shortest distance between two cities, using clearly marked roads. They were then asked to plan a trip route for a person who could only travel a certain distance each day and needed to end in a particular town. The key to doing well on this test was to understand the problem, interpret what was needed, and make connections between reasoning, using representations, and making calculations. Only 12% of U.S. students (compared to an international average of 18%) were able to thoroughly solve multi-step problems like this one. Sadly, 24% of U.S. 15-year-olds were unable even to begin to solve this kind of problem. Overall, U.S. students ranked in the lowest 20% of the countries involved in the study.

The PISA study and others like it call attention to the fact that it is not higher-level math that needs science centers' attention as much as basic problem-solving and solid reasoning. This kind of problem-solving is defined by the PISA report as "understanding information that is given, identifying critical features and any relationships in a situation, constructing or applying one or more external representations, resolving ensuing questions, and finally, evaluating, justifying, and communicating results" (PISA, 2003, p. 24).

Interestingly, these processes are critical to problem-solving in scientific, mathematical, and everyday life contexts. Working on problem-solving in youth and outreach programs could increase participants' mathematical understanding as well as their ability to approach and analyze all kinds of problems.

Youth and outreach programs of science centers are already focusing some attention on problem-solving, albeit primarily in the context of science. In the three examples that follow, we demonstrate how centers have introduced explicit mathematical problem-solving into their existing youth and outreach programs.

In all cases, these centers have engaged students in math that is engaging, relevant, and problem-based. The two centers that work with high-school youth have also made explicit connections among math, service, work, and academic success. The experiences of all three of these centers demonstrate that youth and outreach programs can be ideal settings for teaching and learning mathematical problem-solving.

Saying "YES" to Math at the Saint Louis Sciencenter

The Saint Louis Sciencenter's Youth Exploring Science (YES) program offers its participants, aged 14 and older, opportunities to explore scientific concepts through inquiry-based experiences. YES teens also receive job skills training, as well as opportunities to teach what they are learning to younger children. Many of the youth involved with the project participate over a two-year period, providing opportunities for extended engagement in math and science. The program is operated out of the Taylor Community Science Resource Center, which is equipped with a wet lab, a bank of computers, conference rooms, and plenty of space to conduct projects. An acre plot of land across the street from the Taylor Center, called Science Corner, provides a unique opportunity to explicitly integrate math and science experiences. YES teens plant trees, monitor their progress by taking measurements, and analyze data on the growth and health of the trees. Back in the computer lab, they conduct data analysis with the easy-to-use statistical software, TinkerPlots[1]. Throughout the programs of Science Corner, math is integrated into scientific work.

The directors of the program, Lance Cutter and Nao Ueda, divide the Science Corner work into four strands: Planter Boxes, The Lab, Landscaping, and Technology. Teens choose one strand to stay with throughout a session several weeks long.

Integrating math and science into gardening projects, Saint Louis Sciencecenter.

[1] Published by Key Curriculum Press, Emeryville, CA.

Mathematical problem-solving involving measurement and data were pervasive in all of the strands. For example, in the Planter Box strand, each teen has his or her own plant to nurture and monitor. The goal is to investigate what are the optimum amounts of water and fertilizer for growth. In the service of this goal, there are many opportunities to collect data and practice measurement. For example:

✓ Measure, record, and compare vine length and number of leaves.

✓ Measure, record, and compare trans-leaf area.

✓ Monitor and adjust nitrogen level of the soil.

Throughout their work, the teens learn about measuring variables that are difficult to measure. For example, figuring out the area of a leaf is not the same as figuring out the area of a geometric figure such as a rectangle, as one would do in school. This is particularly true when there are holes and gaps in the leaves, in many cases caused by insect damage. Lance Cutter reflects on the process: "An issue that kept coming up as we worked with the teens was How do you measure and observe something? By observing better, teens were able to notice when insects were eating the plants. By measuring surface area of leaves, they were able to determine the extent of damage caused by insects. They spent a lot of time on basic measurement."

Nao Ueda, Lab Director for Science Corner, shares an example of another non-routine measurement activity. This activity involved designing the space at Science Corner and determining how much to devote to different purposes. "First, we had the teens walk the plot and get a real idea of what 100 ft x 100 ft looked like. We made a to-scale map of the plot from the measurements and drawings the teens made. The teens also did an inventory of the plot and discussed what they already had, what they needed, and how much new things would cost."

The teens marked the perimeter of their garden with spray

Monitoring tree health at Science Corner, Saint Louis Sciencecenter.

paint, according to the plan they had made on paper, and workers came in and tilled the ground for them. They then sprayed the finer details on the tilled ground, such as the footpath, and split the plot up into sections. They figured out where a pond would go in their overall scheme.

Teachable moments emerged, for example, a discussion on how far apart the plants should be and what was the ideal concrete-to-water ratio for this particular project. According to Nao, "We addressed these questions as they came up by gathering more information about the topic via the Internet and by asking questions of experts."

Among the challenges that Lance and Nao faced was the diversity of math backgrounds of the YES teens. Yet, as Nao points out, these challenges can also be thought of as opportunities. "Even if the teens aren't good at math in school, their abstract reasoning skills are becoming more developed, which allows for different opportunities to arrive at the same conclusion. The students who excel in school math can take a formula and run with it, while the ones that struggle may be able to reason through making a mixture without using formal algebraic expression. Because our teens always work in groups, they chose the people in their group who they thought had the skills to do the tasks that used math. I had worried about some teens shrinking from the math and leaving it to others in their groups, but I think once the atmosphere was established, teens would demand that I sit down with them and explain how to solve something until they could get it on their own."

Lance explains that among teens in the YES program, there's typically a split between academically strong participants (those taking geometry, trigonometry, and calculus in school) and those who shy away from rigorous math courses. "The cool thing about Science Corner is that you can be either kind of teen and enter the program."

Learning by Teaching Others

The Science Corner program also has a teaching component in which the YES teens provide free classes to community groups, summer camps, and group homes. The teens are split into three teams of five to six teens each and conduct classes for younger children at the Taylor Center. The teens are trained to prepare, set up, and run their classes with minimal supervision. Although most of their "students" are younger than the teens, throughout the summer they interact with ages ranging from preschoolers to

Collecting and analyzing data from balance challenge, Saint Louis Sciencecenter.

senior citizens. They also work with students from the school for the deaf, people with physical and mental disabilities, and a small number of children who do not speak English. The teaching component provided Nao with another opportunity to introduce mathematical problem-solving. She offers this explanation:

> One of my objectives this summer was to get the teens to start representing number, distance, and other attributes of interest on paper, especially in the form of graphs. I remembered from my time teaching in the public schools that teens who can read graphs and understand their components are miles ahead of their nongraph-savvy peers.

Nao built graphing and data collection activities into the balance challenges she designed for the younger students and presented the activities to her teen team teachers:

> I had the teens go through each challenge and at first let them research the activities for a prolonged period of time. I threw out some questions, such as "Would you balance longer with your eyes fixed to a dot on the wall or, to the tip of someone's finger?" or "Can you balance longer on one foot with your arms out like an airplane or, close to your sides?" I invited the teens to pose some of their own questions.
> I then had the teens pick one question that they wanted to

explore as a group, and we talked about how to make a graph where we could collect our data. What kind of graph would work best for this situation? Our first group question was "Can you get across the tightrope (a thick rope taped to the floor) faster with your arms folded in front of you or out like an airplane?" We made a bar graph with sticky notes and large chart paper, and timed each teen. The teens wrote their names and best times on their sticky notes and stacked them on the "arms folded" side or the "arms out" side. When the teens were ready to teach their classes, they had three graphs and three questions, to correspond to their activities. In addition to teaching the anatomy of balance and the graphing, I was able to work in a few public speaking workshops to build up the teens' confidence through scenarios and role-play activities.

Math Is Authentic Work

Giving teens the task of teaching in science-center programs harkens back to Exploratorium founder Frank Oppenheimer's vision to have high-school students as the museum's primary explainer staff. His intention was to have welcoming nonexperts to convey enthusiasm and openness about what they were learning to visitors they encountered at exhibits. Another important part of this strategy was that the explainers were engaged in what they knew was real work, i.e., activities that the museum depended on them to do. Because the Science Corner classes at the Taylor Center are one of the many community science activities at the Saint Louis Sciencenter, the authenticity of the work is still as much a key point as it was in the early days of the Exploratorium.

Lance Cutter explains, "Initially the teens viewed the authentic work as the shoveling or planting of their sweet potato projects. If it wasn't physical, it wasn't work to them. So Nao and I had to talk to them about how you could be doing 'work' if you were sitting at a table doing a math problem." Nao adds, "We tried to make the program as project-based as possible and learned that as long as the math was necessary to complete a certain task within a larger project, the teens did the math as a matter of course."

Have YES teens become more mathematically capable and confident in these out-of-school hours? Lance offers this observation: "I did have several older teens tell me either that [this work] helped them with their chemistry class or that they wish they had been doing this project while they were taking chemistry classes. I talked to them about this for a while, and I think we made math a little less abstract, and more relevant and user-friendly for them."

Math is TEAMS work at the Lawrence Hall of Science

The TEAMS program (Teens Exploring and Achieving in Math and Science) at the Lawrence Hall of Science (LHS) in Berkeley, California, provides career training, classes, volunteer internships, and work-based positions for teens. Among other activities, students in the program facilitate visitor interactions and perform demonstrations on the floor of the museum.

Working to incorporate more math into their outreach program, LHS engaged TEAMS teens in a popular outdoor exhibition called Forces That Shape the Bay (FSB), one of the areas where the TEAMS teens perform demonstrations. The genesis of the project was a thoughtful math floorwalk of the exhibition undertaken by a group of LHS staff. Through the lens of math, they began to see the possibilities for developmentally appropriate activities that involved measuring volume and area, using a sand exhibit and the irregularly shaped pond that is the centerpiece of the exhibition. The planning process that ensued galvanized a new collaboration between the exhibits department, TEAMS, and the Alliance for Collaborative Changes in School Systems (ACCESS), an LHS math professional development program for teachers.

A collaboration with the South Berkeley YMCA after-school program provided an authentic opportunity in the form of a ready and willing audience with whom the TEAMS teens could prototype the new math programming. The after-school children came to LHS for a special visit involving the FSB Exhibition. Teens were given the goal of developing the math activities for this visit.

ACCESS director Harriette Stevens recalls, "It took many months of walking around with the teens, learning math. During the process we were informally interviewing them. Some teens said that they didn't see any math in the FSB Exhibition. Ultimately, the kids were a big part of making the activities fun. In a training session prior to working with the children, TEAMS teens learned how to use cord or string to estimate the perimeter of an irregularly shaped object, in this case, the pond that is the centerpiece of the FSB Exhibition. After several prototyping sessions, the teens were ready to facilitate these activities with the younger children.

The big mathematical ideas identified for the project with

Measuring the perimeter of a pond, Lawrence Hall of Science.

the after-school children were measurement, scale, and proportional reasoning. These activities were divided into two stations; the instructions are summarized here:

Sand tables in mountain area: volume comparison activity

Children use different-sized containers, such as cylinders and

How many cones in a cylinder?
Lawrence Hall of Science.

cones, and different sifters to explore volume. Older children measure and explore the proportionality of sand to sediment.

In the orientation for the sand activity, the children were first given the opportunity to freely explore the materials—sand, shovels, cones, and cylinders. After identifying the measuring tools (e.g., cone, cylinder), students predicted how many cones of sand it would take to fill a cylinder. Finally they scooped sand into cones to find the answer.

Measure pond: measurement of irregular shapes

TEAMS teens created a laminated drawing of the large pond in the FSB Exhibition, to be used in a game estimating the surface area of irregular shapes. The children then fit rectangles into the shape to estimate the size of the pond's surface area.

On the day of the visit, TEAMS teens and ACCESS staff hosted 35 school-age children and family members from the South Berkeley YMCA (SBY). Located in the lowest income neighborhood of Berkeley, the SBY provides after-school and summer-long educational enrichment programming for its underserved community. Although annual field trips to LHS have been a tradition at the SBY program, participating in teen-facili-

Examining area in "Forces that Shape the Bay" Lawrence Hall of Science.

tated math activities was a new experience. Part of the attraction for SBY director Shirley Brower was the environmental component of the FSB Exhibition. Because the environment has been the focus of other SBY programs, it provided an entry point into the math the children would be exposed to. Ms. Brower was happy with the results. "A great deal of planning went into this. The kids were working kinesthetically with the concepts of volume and density, and they didn't leave feeling stupid. It heightened our expectations of what the Hall [LHS] has to offer our community, and we hope to work in similar ways with them in the future."

The experience was also positive as an outreach program, especially the strategy of teens teaching younger children. Naomi Stein reflects, "Teens are a bridge between children and adults. Their manner is more approachable and they remember more clearly what it's like to be a child, so they are empathetic. They are inherently 'cool' to the [younger] students, and this cachet lends itself to the activities."

LHS' approach to this project demonstrates the best of what it's possible to do with math in an outreach setting. Both of the activities started with play, without formulae, and in the true spirit of inquiry. In school there's often not the time or infrastructure to do the measurement work that involves the messiness of sand or water on the scale of the exhibits in Forces That Shape the Bay. It also allowed for a combination of kinesthetic and mathematical experiences that was valued by the South Berkeley YMCA.

For students whose school experiences are complemented by

programs like this, it provides an opportunity to be successful in math in an informal setting. On a cognitive level, the measurement, scale, and proportional reasoning skills developed in such activities are fundamental to more sophisticated understandings involved in mathematical problem-solving.

One of the outcomes of this project was the cooperation across various LHS programs. The Lawrence Hall of Science has a long history of groundbreaking work in several of its math programs, including ACCESS, EQUALS, and Family Math. Bringing these strengths to teen and outreach programs was a logical next step.

Math, Public Television, and Community Outreach at the North Carolina Museum of Life and Science

Education staff at the North Carolina Museum of Life and Science (NCMLS) discovered a resource from another part of the informal science education world. Public television became the tool that they used to infuse mathematics into an already vital outreach program. Through their existing partnership with the Durham Public Schools, NCMLS decided to introduce *Cyberchase©*, a math-based educational Public Broadcast System program that includes videos and accompanying activities. This program is described in more detail at its Web site, www.pbskidsgo.org/cyberchase.

The *Cyberchase* work was incorporated into the ongoing after-school programming for elementary school students conducted by NCMLS. This mathematical work is added to existing programs for after-school students in animal science, physical science, and chemistry. NCMLS educators have been doing *Cyberchase* programming for the past four years in elementary and middle schools and community centers, as well as at the museum during their popular summer camps.

Far from a passive experience, *Cyberchase* engages the students first with a pre-viewing discussion, followed by an entertaining, suspenseful program and post-viewing activities. The *Cyberchase* Workshops in a Box activities provide viewers with ways to maintain the mathematical momentum generated by the program, while providing activity leaders with easy-to-use games that develop children's mathematical skills.

Cyberchase is an animated show that features a multiethnic cast and mathematical plot lines. The activities include looking closely at shapes in nature, investigating standard units of measurement, and using deductive reasoning to solve "math mysteries." Each TV show is accompanied by a set of activities that can be implemented in after-school settings. According to NCMLS educator Nancy Dragotta-Muhl, "The programs are easy to facilitate and maintain, engaging, and they model diversity."

A snapshot of one program, "Snelfu Snafu" (Part 1), is provided to give readers a sense of ways in which engaging mathematical content can be offered to elementary school children.

Snelfu Snafu, Part 1
The Big Idea: When you save small amounts of money (in denominations of "snelfus" on the program) at a steady rate, your savings will grow larger, and you can predict when you will have the amount you need to buy something you want.

Math Content: Number & Operations
Episode Description
Motherboard, The CyberSquad (Inez, Jackie, Matt, and Digit), and Dr. Marbles are all good characters on the program, and Hacker, Delete, and Buzz are the bad guys. The show uses the tension of plots like the one described below to sustain children's interest.

The long lost encryptor chip, the only computer component that can restore Motherboard to full power, unexpectedly appears on the popular cyberauction U-WANT, U-BID! To make matters worse, Hacker is the top bidder—and if he gets the chip, Motherboard is doomed! The CyberSquad and their friend Slider get jobs in Cyberspace and figure out how much they need to save per day to outbid Hacker. They succeed in winning the auction but are totally unprepared for what happens when they install the chip in Motherboard... Hacker takes over Cyberspace!

NCLMS staff worked with second- and third-graders at Glenn Elementary School, using a series of *Cyberchase* episodes such as the one described above. Glenn Elementary is in a rural part of Durham, where the population has a broad mix of socioeconomic and ethnic backgrounds. The participants in the program come directly to the NCMLS/*Cyberchase* program after school. They typically have a snack and guided homework or tutoring time before making choices for group enrichment activities such as *Cyberchase*.

Pre-episode activities typically include a discussion like the one that NCMLS Programs Manager Leiana Leon Guerrero led around the "Snelfu Snafu" episode. She began by asking the children whether they had ever saved money for something they really wanted. She then asked them how they earned the money, what their process was for saving, and whether or not they achieved their goals. The children mentioned wanting skateboards, Game Boys®, and a (fake) tattoo. They said that they have earned money by doing chores and homework, and by playing well at baseball.

After watching the episode, Leiana gave instructions for an activity in which the children perused newspaper ads, identified what they wanted to buy, and came up with savings plans. She told the children that they needed three pieces of information to figure out how long it would take to save the right amount of money: How much does the item I want to buy cost? How much do I earn (per job or hour)? and How frequently do I make this money?

In a follow-up activity, Leiana and three other educators worked with the children to create spending plans for clothing, iPods®, video games and systems, and cars. Some children even aimed at saving for home appliances such as refrigerators and washer/dryer combinations. They talked about saving the money in weekly increments—mostly through chores or odd jobs they could do for family members or neighbors. Staff encouraged them to take their plans home and work toward their goals with their parents' support and supervision. NCLMS staff reported that the students' level of engagement is high and that they have had good feedback from some of the children's teachers. For example, one teacher from Glenn Elementary wrote the following:

> My second- and third-graders loved using the *Cyberchase* programming as part of our after-school curriculum. The children quickly identified with the characters, found the segments engaging and entertaining, and even sang along with the theme song! Many of them mentioned watching *Cyberchase* on PBS at home and expressed that doing the activities was "a fun treat… not work."

The episodes work well for all levels of learners. Some of the more advanced children quickly see the parallels between the problems presented in the activities with the problems the *Cyberchase* characters face, and figure out solutions by using the strategies they saw characters employ. Others use their own strategies for solving problems.

The program often takes problem-solving to a higher level through extension activities. The "Snelfu Snafu" episode, like all *Cyberchase* programs, comes with accompanying activities and board games. In this case, the game "Gollywood Squares"[2] reinforces the concept of saving money over a period of time.

NCMLS Youth Programs Educator Courtney Pickett states that the one of the keys to success with second- and third-graders is alternating the programs with other activities that are also part of the museum's outreach program, so that children have a variety of formats for after-school learning. In the future, she and the other Youth Programs staff would like to train teens in the NCMLS Youth Partners Program to lead *Cyberchase* sessions. From an outreach perspective, the power of the program is that it provides an accessible and low-cost way for science centers to involve their after-school community partners in math.

It is worth noting that another institution involved in the Math Momentum project, the Children's Museum of Houston, has used the *Cyberchase* TV program in a quite different way, as the foundation for a new exhibition tentatively called Cyberchase: The Chase is On! Using pieces from episodes as prototypes, and with attention given to strands of the National Council of Teachers of Mathematics Standards, this 2006 exhibition will be accompanied by 120 activities and 15 kits. In both North Carolina and Houston, the power of the *Cyberchase* program is the use of a highly motivating fictional context that draws children into doing meaningful mathematics.

Conclusions and Next Steps

In all three of the examples in this chapter, math is incorporated into outreach programs as an extension of existing work, rather than as separate "math programs." With this approach there is less chance of raising math-phobia flags for audiences who may choose not to visit an exhibition or participate in a program focused solely on math. At the same time, it is important to note that math is not hidden from participants; rather, it is a natural ingredient of the programs. In St. Louis and Berkeley, teens who entered the program as a means of working with science found that math is a necessary part of science. Explaining science and leading science activities inherently involves knowing math.

The model of involving youth as "math ambassadors" is sim-

[2] http://pbskids.org/cyberchase

ilar to work that has been done in science, particularly through the YouthALIVE![3] initiative. A recent article of the American Association of Museums publication *Museum News* points out, teens "bring with them a highly developed curiosity and a willingness to experience new ideas in an environment quite distinct from home or school" (*Museum News*, 2005). One important goal of this approach is the academic enrichment of youth. As can be seen in the Lawrence Hall of Science and Saint Louis Sciencenter examples, young people need to understand mathematics and its uses (in these cases, its scientific uses) in order to be effective leaders of activities. The work involves *doing* mathematical problem-solving for a purpose, rather than simply leading younger students through a script. A great deal of mathematical preparation is necessary before teens work with younger students.

One need in youth programs involving math is more evidence about the impact of programs on participants' learning. Because these programs typically involve youth for a relatively long period of time—in most cases, one or more years—the question of impact naturally arises. Programs such as YouthALIVE![3] provide evidence that participants' attitudes about school have become more positive, especially as they see the relevance of science to their current and future lives. These programs "are providing youth with meaningful and authentic experiences. The strongest and most common areas of benefit for participants are the development of their communication skills, a shift in their attitude toward learning and an increase in personal self-esteem… Over half the youth surveyed say that their experience has strongly encouraged them to continue in school" (Inverness Research Associates, 1995, p. 5).

These results, coupled with more recent experiences like those described in this chapter, are extremely encouraging. They suggest that the next step is to focus on the mathematics experiences incorporated into the youth and outreach programs offered by science centers. If math-oriented youth and outreach programs are successful in engaging students in mathematically relevant peer teaching, service, and community work, we hope to see the following outcomes:

✓ Increased ability to describe the uses of math in science

✓ Increased understanding of the role of math in potential careers

✓ Increased confidence in ability to solve real-life problems that involve mathematics

[3]YouthALIVE!—Youth Achievement through Learning, Involvement, Volunteering, and Employment—is an initiative that was supported by the DeWitt-Wallace Reader's Digest Fund and administered by ASTC from 1992 to 2002 to increase the capacity of science centers and museums to develop programs for and with adolescents, and to enhance the ability of museum staff to work with youth.

Those who are considering mathematizing youth or outreach programs have an opportunity to contribute to a much-needed research base on informal math learning. There is a huge gap in our understanding of out-of-school math learning, and it is incumbent upon those who are doing youth and outreach programs to begin addressing this gap. One logical starting place is to involve youth in creative, activity-based assessments of mathematical problem-solving. For example, simply asking participants to engage in a real measurement activity, such as figuring out the dimensions of their meeting space, provides important information on skills. Done on a pre- and post-assessment basis, this type of activity could provide important data on program impact.

References

Dude, where's my museum? 2005. *Museum News.* September/October, 84:5.

Inverness Research Associates. 1995. Executive summary. YouthALIVE!: A review of the program's first three years. Unpublished report prepared for the DeWitt Wallace Reader's Digest Fund. New York, NY.

Organisation for Economic Co-operation and Development, Programme for International Student Assessment (PISA). 2003. *Problem Solving for Tomorrow's World.* 1–154.

Konold, Cliff. Tinkerplots: Dynamic Data™Exploration. Software for Grades 4–8. Emerville, CA: Key Curriculum Press.

Sutherland, C. 2006. Joining the *Cyberchase:* A cross-platform math opportunity. ASTC Dimensions. January/February:18.

Turner, J., M. Gheen, E. Anderman, and Y. Kang. 2002. The classroom environment and students' reports of avoidance strategies in mathematics: A multimethod study. *Journal of Educational Psychology* 94:88-106.

Appendix

Math Projects from Thirteen Science Centers

Note: These projects are representative of the ones undertaken for the Math Momentum in Science Centers Project (MMSC).

Buffalo Museum of Science

1020 Humboldt Parkway
Buffalo, NY 14211
Tel: (716) 896-5200
Web: http://www.sciencebuff.org/
Project Contact: Jayme Cellitioci

Math Project: Mathtodons & Co.

Through Mathtodons & Co., the Buffalo Museum of Science's MMSC project, we have increased the use of mathematics in a paleontology prototyping gallery focusing on the Museum's Ice Age dig site. The result is a multigenerational experience for the general public in our Byron Dig Experience Lab gallery. Additionally, we have incorporated math activities into kits (for children aged 5–12) distributed to 12 community center partners associated with the museum. The major math concepts that we have focused on are measurement, data, graphing, and ratios.

Even prior to the start of this project, it was a goal of ours to increase public knowledge of our Ice Age dig site research, so this was a great opportunity to double prototype! We began by identifying the various ways that math is used in paleontology in the field and lab. As we learned about children's needs for more informal measurement and math experiences, we also focused our efforts on enhancing the kits for the community centers.

Through this project, we have strengthened our ties with our visitors and community groups, as well as enhanced collaborations between the Museum Paleontologist, college interns, TERC staff, and our community center outreach staff.

Children's Museum of Houston

1500 Binz
Houston, TX 77004
Tel: (713) 522-1138
Web: http://www.cmhouston.org/
Project Contact: Keith Ostfeld

Math Project: Bubble Math

The Bubble Math project at the Children's Museum of Houston engages families in activities that focus on geometry, measurement, data collection, and problem-solving with bubbles. The project core is ten bilingual kits that are facilitated by a cadre of Museum staff in the exhibits and at family learning events hosted at elementary schools. They are also available for checkout through the Museum's Parent Resource Library (a branch of the Houston Public Library system) and will soon be included in the *Magnificent Math Moments Family Learning Guide*, a four-volume set of 120 math activities that parents can do with their children at home.

Bubble Math was developed in partnership with Dr. Jan Mokros as an extension of the Museum's Bubble Lab exhibit activities. The project is aligned with the Museum-wide DEEP Initiative, which works to ensure that the concepts explored in exhibit investigations are solidified through a variety of other related activities.

Fort Worth Museum of Science and History

1501 Montgomery Street
Fort Worth, TX 76107
Tel: (817) 255-9300
Web: http://www.fwmuseum.org/
Project Contact: Colleen Blair

Math Project: Lone Star Dinosaur Family Activity Guide

The Lone Star Dinosaur Family Activity Guide, a four-page, bilingual math activity take-home collateral, targets families with children aged 5–11. The guide is designed to support and facilitate families' post-visit conversations about dinosaurs. The big mathematical idea is measurement. The activity guide is one of five educational outreach projects supporting the museum's new 8,000-square-foot exhibition, Lone Star Dinosaurs. The exhibition, collateral materials, and outreach programs were funded in part by the National Science Foundation. It is important to note that the exhibition introduces five new Texas dinosaurs that have been discovered within the past 18 years. Two of the dinosaurs were actually found by children.

The intent of our project was to build on children's natural curiosity about dinosaurs and create simple measuring activities involving everyday objects found in their homes and backyards. Original plans for the guide focused purely on science process skills. As our staff's participation in MMSC's professional development and workshop experiences continued, our vision for the guide shifted to include mathematics. Our participation with the MMSC enabled us to extrovert the mathematics that was naturally embedded within our new exhibition.

Lawrence Hall of Science

University of California, Berkeley
Berkeley, CA 94720
Tel: (510) 642-5132
Web: http://www.lawrencehallofscience.org/
Project Contact: José Franco

Math Project: Forces That Shape the Bay Activities

The Lawrence Hall of Science's MMSC team developed a set of mathematics activities for an existing outdoor exhibit, Forces That Shape the Bay (FSB). The focus on measurement, proportionality, and proportional reasoning addresses the National Council of Mathematics Teachers Standards and complements the mathematics in Forces. The activities are inquiry-based, hands-on, and include play with measurement apparatus. Visual, auditory, and tactile experiences address diverse skill levels.

We began by consulting with LHS staffs from professional development, exhibits, and Teens Exploring and Achieving in Math and Sciences (TEAMS), and then conducted a walk-through of Forces. The group identified ways to emphasize the mathematics and contacted a local community-based organization, South Berkeley YMCA, which reflects the ethnic and economic diversity of the broader community. During the development process, teens from our TEAMS program were also trained to lead these activities. The resources will enrich a permanent exhibit and help address the needs of teachers, students, and their families by offering quality programming. We will use the process as a staff-training model and add the activities to museum programs.

Miami Museum of Science and Planetarium

3280 South Miami Avenue
Miami, FL 33129
Tel: (305) 646-4200
Web: http://www.miamisci.org/index.html
Project Contact: Cheryl Juarez

Math Project: Ruling Reptiles!

The Miami Museum of Science and Planetarium developed plans for a new reptile program as part of their MMSC work. In the Ruling Reptiles program, visitors will explore how a real scientist uses measurement and data collection in crocodilian research and how it relates to the conservation of the three species that live in the Florida Everglades. Principal themes include why crocodiles are ecologically important; how the accumulation of biological data (capture, measure, tag, release) helps to ensure their survival; and how scale, ratio, and perception are important to measurement. The program will be delivered by the Museum's Wildlife Educators and will feature hands-on interactive components, as well as high- and low-tech equipment that will encourage visitors to measure, extrapolate lengths, and compare body ratios.

Museum staff representing exhibits, education, and youth programs worked together to develop the Ruling Reptiles concept in collaboration with their wildlife center staff and a local scientist. The Wildlife Center was selected for the program because of the numerous ways measurement is used in the real world of animal care, research, and conservation. The overall goal of the project is to help visitors of all ages and demographics better understand concepts involved with measurement, primarily scale, ratio, and proportion.

Museum of Life and Science, Durham, NC

433 Murray Avenue
Durham, NC 27704
Tel: (919) 220-5429
Web: http://www.ncmls.org/
Project Contact: Nancy Dragotta-Muhl

Math Project: *Cyberchase*[©] Program

The Museum of Life and Science has worked on a new program based on the popular PBS children's series *Cyberchase* as part of its MMSC commitment. The program incorporates use of the TV-series footage and the companion Workshops in a Box, for sharing math discovery at after-school program sites, 21st Century Learning Centers, and on site at the Museum of Life and Science. The target group we identified was students that we serve in our after-school programming at high-priority schools.

Cyberchase was a great fit for the programming we do and the audiences we currently serve. The program is dynamic, accessible, and models good diversity. It is an engaging way to help encourage interest in math, facilitate problem solving, strengthen content knowledge, and narrow various equity gaps that contribute to disparities in math performance. We got started with the program after being introduced to it during a workshop by Channel 13, New York. We continue to work with them as they produce new resources surrounding the show.

Museum of Science, Boston, MA

Science Park
Boston, MA 02114
Tel: (617) 723-2500
Web: http://www.mos.org/
Project Contact: Loren Stolow

Math Project: Math in Museum of Science Exhibits: Resources for Educators

The Museum's MMSC Team developed Mathematical Questions & Things to Do for four permanent exhibits to help teachers and learners draw out the math. These questions are mapped to the National Council of Teachers of Mathematics standards and are available as part of a larger Museum of Science Web site database of resources for educators.

These resources will increase accessibility of mathematical ideas such as measurement, data analysis, probability, and algebra to school-group learners by providing tools to their teachers and chaperones.

We began by brainstorming with fellow Museum staff members, which led us to the larger resource database project. Joining this project in its early stages afforded us the opportunity to include math. We were able to discover that highlighting the extensive math already in our exhibits is as valuable as creating new work.

New England Aquarium

Central Wharf
Boston, MA 02110
Tel: (617) 973-5200
Web: http://www.neaq.org/
Project Contact: Rebekah Stendahl

Math Project: Penguins

As part of the MMSC project, the New England Aquarium focused on a classroom program called Penguins. The Penguin program takes place in classrooms at the Aquarium and at schools and community centers throughout the region. The target age is second through fourth grades, although we have adapted the program for both older and younger audiences. The class focuses on data collection and analysis.

We knew that penguins were an animal we could have in our collection for years to come, in addition to being a visitor favorite. After initial consultations with our penguin husbandry staff, we were able to choose an appropriate focus for data collection. Through this program, we hope to enhance a participant's experience and create a stronger connection with the natural world by encouraging the study of individual penguins as a way to learn about the colony as a whole. We believe that this will lead to a desire to protect aquatic environments, thus furthering the institution's mission to present, promote, and protect the world of water.

New Jersey Academy for Aquatic Sciences

1 Riverside Drive
Camden, NJ 08103
Tel: (856) 365-0352
Web: http://www.njaquarium.org
Project Contact: Angela Wenger

Math Project: "How Much Food Would You Eat If...."

The New Jersey Academy for Aquatic Science's MMSC project centers on comparing the food consumption rates of Sandtiger sharks and African penguins. In this engaging, facilitated floor program, a guest is invited to weigh in on a bathroom scale to determine his or her weight in pounds, and then plot the weight on two separate graphs that represent the amount of food that a similarly sized shark or penguin would consume daily. Sandtiger sharks typically eat 2–3% of their body weight each week, while a penguin will eat as much as 20% of its body weight every day.

Pounds, as a unit of measurement, are often difficult to visualize. Therefore, guests stack an appropriate amount of life-size, food-shaped boxes to come to an understanding of how much the various animals each eat. For example, a 100-pound child will see that if he or she were a penguin, he or she would eat the equivalent of a 20-pound Thanksgiving turkey every day, while a shark the same size would only eat the equivalent of a few slices of cheese each day. Math concepts that this project touches on include working with graphs, greater than/less than values, weight measurements, and percentage values.

Oregon Museum of Science and Industry

1945 S.E. Water Avenue
Portland, OR 97214
Tel: (503) 797-OMSI (6674)
Web: http://www.omsi.edu/
Project Contact: Marilyn Johnson

Math Project: ECE Math Discovery Boxes and Drawers

OMSI's MMSC project incorporated mathematical materials into OMSI's early childhood Discovery Boxes and smaller Discovery Drawers. The boxes consist of introductory science-based curriculum integrated with literacy, math, art, music, dramatic play, sensory skills, and motor skills. They are available to educators and volunteers throughout the museum to enhance exhibits and provide math activities for preschool children. Parents and children practice concepts using Discovery Drawers at their own pace, giving parents the tools to be their children's first and best teachers. Parents are impressed with their children's ability to be interested in and understand math.

Funding and professional development from the TERC/ASTC MMSC project enabled OMSI to build a Weights & Measurement Discovery Box and four smaller Discovery Drawers (Weights & Measurements, Numbers & Counting, Shapes & Patterns, and Telling Time) for our early childhood lab. These drawers include instructions for educators and parents, and such tools as measuring tapes, a balance scale, and volume measures for exploration and experimentation.

Saint Louis Sciencenter

5050 Oakland Ave.
St. Louis, MO 63110
Tel: (314) 289-4400
Web: www.slsc.org
Project Contact: Diane Miller

Math Project: Science Corner

Science Corner was an existing component of the Youth Exploring Science program that was created to teach teens about matter and energy. Because this section was geared towards training teens in lab skills, it was an ideal forum to study the integration of math into a science program.

When the teens were engaged in their plant-cloning project, they not only learned proper lab techniques, but also how to use math concepts such as formulas and ratios to make the media. This project allowed the teens to look at math as a means to completing their projects, rather than doing math for the sake of manipulating numbers. Teens in Science Corner worked on these projects in groups of individuals with different math abilities. The challenge for the staff was to provide a math- and science-rich program in which all the participants could feel successful.

Science Museum of Minnesota

120 West Kellogg Blvd.
St. Paul, MN 55102
Tel: (651) 221-9444
Web: http://www.smm.org/
Project Contact: Maija Sedzielarz

Math Project: MathPacks

SMM developed MathPacks, a non-staff-facilitated school program including tool-filled backpacks to be used in museum galleries, supported by online background information and pre- and post-trip classroom activities, and aligned with state science and math standards. MathPacks are also available to other groups, such as community partners and Youth and Family workshops. The big mathematical ideas introduced include measurement, mathematical relationships, meaning of arithmetic operations, and ratio.

The first MathPack focus is Measuring Growth, funded through the Medtronic Foundation's STAR Grant program. This focus was developed by interviewing SMM scientists about math